Matt Jackson

Part 2 work
O'K
M.J.

MW00713850

Laboratory Exercises

to accompany

Introduction to
Mechatronics and
Measurement Systems

Second Edition

David G. Alciatore
and
Michael B. Histand

Department of Mehanical Engineering
Colorado State University
Fort Collins, CO 80523

Refer to the textbook website for more information:
www.engr.colostate.edu/mechatronics

Boston   Burr Ridge, IL   Dubuque, IA   Madison, WI   New York   San Francisco   St. Louis
Bangkok   Bogotá   Caracas   Kuala Lumpur   Lisbon   London   Madrid   Mexico City
Milan   Montreal   New Delhi   Santiago   Seoul   Singapore   Sydney   Taipei   Toronto

The McGraw-Hill Companies

Laboratory Exercises to accompany
INTRODUCTION TO MECHATRONICS AND MEASUREMENT SYSTEMS,
SECOND EDITION
David G. Alciatore and Michael B. Histand

Published by McGraw-Hill Higher Education, an imprint of The McGraw-Hill Companies, Inc.,
1221 Avenue of the Americas, New York, NY 10020. Copyright © 2004 by The McGraw-Hill
Companies, Inc. All rights reserved.

No part of this publication may be reproduced or distributed in any form or by any means, or
stored in a database or retrieval system, without the prior written consent of the authors and
The McGraw-Hill Companies, Inc., including, but not limited to, network or other electronic
storage or transmission, or broadcast for distance learning.

This book is printed on acid-free paper.

2 3 4 5 6 7 8 9 0 QSR QSR 0 9 8 7 6 5

ISBN-13: 978-0-07-297875-9
ISBN-10: 0-07-297875-9

www.mhhe.com

Table of Contents

General Equipment and Supplies List

Recommended Equipment and Software:

- HP 54602A Oscilloscope
- HP 6235A Triple Output Power Supply
- Philips PM5193 Programmable Synthesizer/Function Generator
- HP 34401A Digital Multimeter
- Microchip PICSTART Plus development programmer
- Microchip's MPLAB integrated development environment windows software
- MicroEngineering Labs' PicBasic Pro compiler

Recommended Supplies:

For each work station (in the station containers):
- electronic components (the required components are listed at the beginning of each laboratory exercise
- alligator clips (4)
- BNC-to-banana connectors (2)
- breadboard (1)
- wire strippers (1)
- static discharge pad (1)

Available for the entire laboratory (hanging on the wall):
- banana cables assorted colors 24inch (32)
- banana cables black and red 24 inch (16 each)
- banana cables assorted colors 48inch (16)
- DMM probes black and red (16 each)
- oscilloscope probes (16)
- assorted BNC-to-BNC cables

Other:
- assorted colors 24 gage solid core wire (100 feet each)
- soldering stations (4)
- solder and flux
- extra soldering tips
- solder suckers/de-solderers

Instrumentation Used in the Lab:

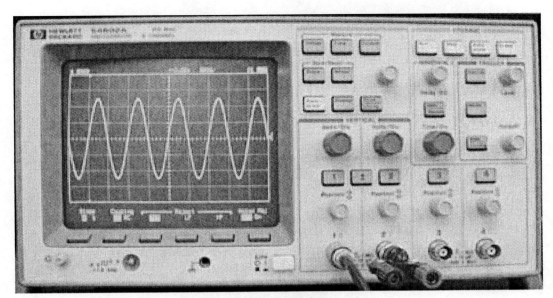

HP 54602A Oscilloscope

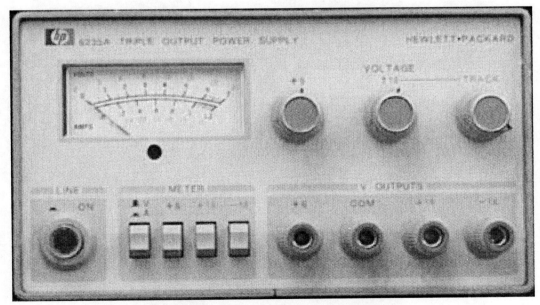

HP 6235A Triple Output Power Supply

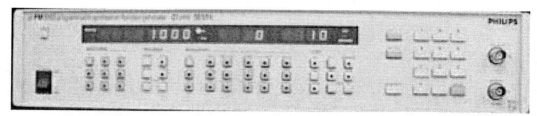

Philips PM5193 Programmable Synthesizer/Function Generator

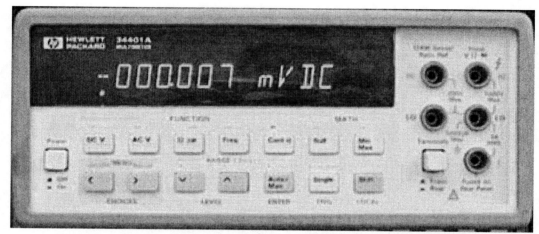

HP 34401A Digital Multimeter

3 labs 41.5
circuit? ∅
total.
41.5/100

Laboratory 1

Introduction - Resistor Codes, Breadboard, and Basic Measurements

Required Components:
- 3 1kΩ resistors

1.1   Introduction and Objectives

Welcome to the world of mechatronics. Your experiences in this laboratory will provide a solid foundation in instrumentation and modern electronics. The purpose of the first laboratory exercise is to familiarize you with the laboratory facilities and procedures, and with basic measurement techniques. The specific objectives are:

- Observe demonstrations of the instruments that you will use throughout the semester. These instruments include the oscilloscope, digital multimeter, power supply, and function generator.

- Learn how to construct basic electrical circuits using a breadboard.

- Learn how to properly make voltage and current measurements in circuits.

- Learn the resistor color code scheme necessary to read resistor values and tolerances.

- Learn about the types of capacitors and how to read their values.

1.2   Electrical Safety

Electrical voltages and currents can be dangerous if they occur at values that interfere with physiological functions. All of the laboratory exercises described in this manual are designed to use ac and dc voltages whose values are less than 15 V, values that will not cause perceptible shock via the skin. If working with voltages higher than these, especially line voltages (110 V_{rms} or 220 V_{rms}), one must be extremely careful to avoid shock or potentially lethal situations. We caution the user of household voltages and currents to carefully read the electrical safety precautions outlined in the textbook (see Section 2.10.1).

1.3   Resistor Color Codes

The most common electrical component found in almost every electrical circuit is a resistor. The type we will use in the Lab is the 1/4 watt axial-lead resistor. A resistor's value and tolerance are usually coded with four colored bands (*a, b, c, tol*) as illustrated in Figure 1.1. The colors used for bands are listed with their respective values in Table 1.1. A resistor's value and tolerance are expressed as

$$R = ab \times 10^c \pm \text{tol} \%$$ (1.1)

where the *a* band represents the tens digit, the *b* band represents the ones digit, the *c* band represents the power of 10, and the *tol* band represents the tolerance or uncertainty as a percentage of the coded resistance value. The set of standard values for the first two digits are: 10, 11, 12, 13, 14, 15, 16, 18, 20, 22, 24, 27, 30, 33, 36, 39, 43, 47, 51, 56, 62, 68, 75, 82, and 91.

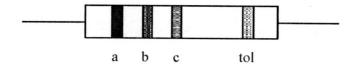

$$\begin{matrix} a & b & c & \text{tol} \end{matrix}$$

Figure 1.1 Wire lead resistor color bands

Table 1.1 Resistor color band codes

a, b, and *c* Bands		*tol* Band	
Color	**Value**	**Color**	**Value**
Black	0	Gold	±5%
Brown	1	Silver	±10%
Red	2	Nothing	±20%
Orange	3		
Yellow	4		
Green	5		
Blue	6		
Violet	7		
Gray	8		
White	9		

1.4   Reading Capacitor values

Most students learn to read resistor values quite easily. However, they often have more trouble picking out a specific capacitor. That's not their fault. They have trouble, as you will agree when you have finished reading this, because the capacitor manufacturers don't want them to be able to read cap values. ("Cap" is shorthand for "capacitor," as you probably know.) The cap markings have been designed by an intergalactic committee to be nearly unintelligible. With a few hints, however, you can learn to read cap markings, despite the manufacturers' efforts to prevent this. Some hints for various size capacitors follow.

Big Capacitors

Big Caps are usually electrolytics. These are easy to read, because there is room to print the value on the cap, including units. You need only have the common sense to assume that, for example, +500MF means 500 micro farads, with the plus indicating the positive end of the capacitor. Be careful to not take the capital M seriously. (Remember the SI system of units?)

All of these big caps are polarized. That means the capacitor's innards are not symmetrical, and that you may destroy the cap if you apply the wrong polarity to the terminals: the terminal marked + must be at least as positive as the other terminal. (Sometimes, violating this rule will form gas that makes the cap blow up; more often, the cap will short internally.

Smaller Capacitors

As the caps get smaller, the difficulty in reading their markings gets steadily worse. Tantalum caps are silver colored cylinders. They are polarized: a + mark and a metal nipple mark the positive end. Their markings may say something like +4R7µ. That also means pretty much what it says, if you know that the "R" marks the decimal place: it's a 4.7 µF cap.

The same cap could also be marked +475K. Here you encounter your first challenge, but also the first appearance of an orderly scheme for labeling caps, a scheme that would be helpful if it were used more widely. The challenge is to resist the plausible assumption that "K" means "kilo." It does not; it is not a unit marking, but a tolerance notation (it means ± 10%). (Wasn't that nasty of the labelers to choose "K?" Guess what's another favorite letter for tolerance. That's right: M. Pretty mean!) The orderly labeling here mimics the resistor codes: 475 means 47 times ten to the fifth power. But what are the Units? 10^5 what? 10^5 of something small. You will meet this dilemma repeatedly, and you must resolve it by relying on the following intuitive observations:

1. The only units commonly used in this country are

   microfarads: 10^{-6} Farad

   picofarads: 10^{-12} Farad

   (you should, therefore, avoid using "mF" and "nF" yourself.)

   A Farad is a humongous unit. The biggest cap you will use in this course is 500 µF. It is physically large (we do keep 1F caps around, but only for our freak show). Thus, if you find a small cap labeled "470," you know it is 470pF.

2. A picofarad is a tiny unit. You will not see a cap as small as 1 pF in this course. So, if you find a cap appearing to claim that it is a fraction of some unprinted unit – say, ".01" – the unit is µF: ".01" means 0.01 µF.

3. A picofarad is not just a bit smaller than a microfarad. A pF is not 10^{-9}F (10^{-3} µF); instead, it is 10^{-12} F: a million times smaller than a microfarad!

So, we conclude, a cap labeled "475" must be 4.7 x 10^6 (47 x 10^5) picofarads. That, you will recognize, is a roundabout way to say 4.7 x 10^{-6} F or 4.7 µF.

We knew that this was the answer, before we started this last decoding effort. This way of labeling is quite roundabout, but at least it is unambiguous. It would be nice to see it used more widely. You will see another example of this exponential labeling in the case of the CK05 ceramic caps, below.

Mylar caps are yellow cylinders, that are rather clearly marked. ".01M" is just 0.01 μF, of course; and ".1 MFD" is not a tenth of a megafarad. You can orient them at random in your circuits. Because they are fabricated as long coils of metal foil (separated by a thin dielectric - the "mylar" that gives them their name), mylar caps must betray their function at very high frequencies: that is, they begin to behave as inductors instead, blocking the very high frequencies they ought to pass. Ceramics (below) do better in this respect, although they are very poor in other characteristics.

Ceramic caps are little orange pancakes. Because of this shape (in contrast to the coil format hidden within the tubular shape of mylars) they act like capacitors even at high frequencies. The trick, in reading these, is to reject the markings that should not be interpreted as units. For example, a ceramic disk cap labeled by "Z5U .02M 1kV" is a 0.02 μF cap with a maximum voltage rating of 1kV. The M is a tolerance marking, in this case (see below), ±20%.

CK05 caps are little boxes, with their leads 0.2" apart so they can be easily inserted in protoboards (AKA perf boards or vector boards) or PC boards. Therefore, they are common and useful. An example marking is 101k. This is the neat resistor-like marking. This one is 100 pF (10×10^1 pF).

Tolerance Codes

Finally, just to be thorough, and because this information is hard to come by, let's list all the tolerance codes. These apply to both capacitors and resistors; the tight tolerances are relevant only to resistors; the strangely asymmetric tolerance is used only for capacitors.

Tolerance Code	Meaning
Z	+80%,-20%
M	±20%
K	± 10%
J	±5%
G	±2%
F	±1%
D	±0.5%
C	±0.25%
B	± 0.1%
A	± 0.005
Z	± 0.025 (precision resistors; context will show the asymmetric cap cap tolerance "Z" makes no sense here)
N	±0.02%

1.5   The Breadboard

A breadboard is a convenient device for prototyping electrical and electronic circuits in a form that can be easily tested and changed. Figure 1.2 illustrates a typical breadboard layout consisting of a rectangular matrix of insertion points spaced 0.1 in apart. As shown in the figure, each column a through e and f through j is internally connected respectively. The + and - rows that lie along the top and bottom edges of the breadboard are also internally connected to provide convenient DC voltage and ground busses for connecting to specific insertion points. As illustrated in the figure, integrated circuits (IC) are usually inserted across the gap between columns a through e and f through j. A 14-pin dual in-line package (DIP) IC is shown here. When the IC is placed across the gap, each pin of the IC is connected to a separate numbered column, making it easy to make connections to and from the IC. The figure also shows an example of how to construct a simple resistor circuit. The schematic for this circuit is shown in Figure 1.3. Figure 1.4 shows an example of a wired breadboard including resistors, and integrated circuit, and a push-button switch. When constructing such circuits, care should be executed in trimming leads so the components lie on top of the breadboard in an organized geometric pattern.

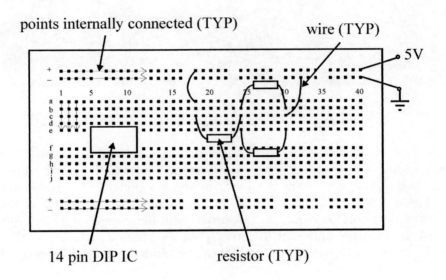

Figure 1.2 Breadboard

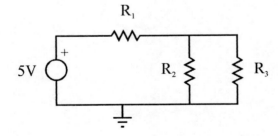

Figure 1.3 Example resistor circuit schematic

11

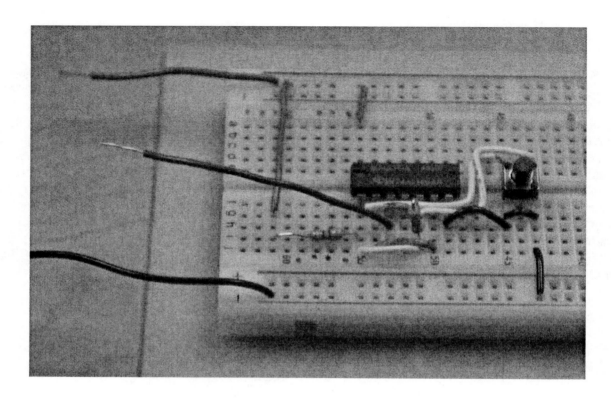

Figure 1.4 Example Breadboard Circuit

It is very important that you know how to make voltage and current measurements, especially when prototyping a circuit. Figure 1.5 illustrates how each measurement is performed. When making a voltage measurement, the leads of the voltmeter are simply placed across the element for which you desire the voltage. However, when making a current measurement through an element, the ammeter must be connected in series with the element. This requires physically altering the circuit to insert the ammeter in series. For the example in the figure, the top lead of resistor R_3 must be removed from the breadboard to make the connection through the ammeter.

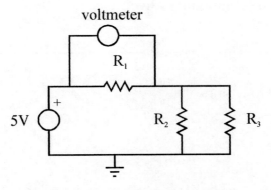

(a) measuring the voltage across R_1

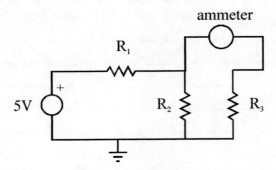

(b) measuring the current through R_3

Figure 1.5 Voltage and current measurements

1.6 Laboratory Procedure / Summary Sheet

Group: _____ Names: _____

(1) Select three 1kΩ resistors.

(2) To verify the reported nominal value, record the color band colors and associated band values, and calculate the nominal resistance from the band colors:

band	color	value
a	brown	1
b	black	0
c	red	2
tol	gold	±5

specified nominal resistance and tolerance:

R = _____1kΩ_____ ± __5%__

(3) Measure the resistance of each of the three resistors provided using the digital multimeter and compare the values with the specified value.

Resistor	Measured Value (Ω)	% Error
R_1	.9860	
R_2	.9890	
R_3	.9777	

(4)   Build the circuit shown in Figure 1.6 with the three given resistors on the breadboard. Note that R_1 is in series with the parallel combination of R_2 and R_3.

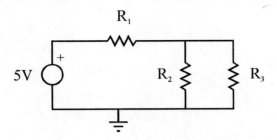

Figure 1.6 Resistor circuit schematic

(5)   Calculate the values for the voltage drop across R_1 and the current through R_3 assuming that all three resistors have equal value 1kΩ. Refer to the text book and Section 2.2 in the next laboratory exercise for background theory. Use the digital multimeter to measure the actual voltage and current values. Refer to Figure 1.5 for the appropriate technique.

	calculated	measured
V_1	3.3 v	3.39 v
I_3	1.66 mA	1.728 mA

If your measured values differ from your calculated values, explain the differences.

$$V_i = I R_1 + I R_e$$

$$\frac{1}{R_e} = \frac{1}{R_2} + \frac{1}{R_3} = \frac{1}{1k\Omega} + \frac{1}{1k\Omega} = \frac{2}{1k\Omega}$$

$$5v = I(1k\Omega) + I(.5k\Omega)$$

$$R_e = .5 k\Omega$$

$$I = 3.3 mA$$

$$V_1 = I R_1 = (3.3 mA)(1k\Omega) = 3.3 v$$

$$I_3 = \frac{1}{2} I_1 = 1.66 mA$$

Laboratory 2

Instrument Familiarization and Basic Electrical Relations

Required Components:
- 2 1kΩ resistors
- 2 1MΩ resistors
- 1 2kΩ resistor

2.1   Objectives

This exercise is designed to acquaint you with the following laboratory instruments which will be used throughout the semester:

- The Oscilloscope

- The Digital Multimeter (DMM)

- The Triple Output DC power Supply

- The AC Function Generator

During the course of this laboratory exercise you should also obtain a thorough working knowledge of the following electrical relations:

- Series and Parallel Equivalent Resistance

- Kirchoff's Current Law (KCL)

- Kirchoff's Voltage Law (KVL)

- Ohm's Law

- The Voltage Divider Rule

- The Current Divider Rule

The experiments to be performed during this laboratory are also designed to introduce you to two very important instrument characteristics:

- The output impedance of a real source

- The input impedance of a real instrument

2.2   Introduction

A thorough explanation of the proper use of each of the instruments above will be presented when you come to the laboratory. You should already be familiar with the basic electrical relations listed above; however, a quick review will follow.

2.2.1   Series and Parallel Equivalent Resistance

It can be shown that when resistors are connected in series the equivalent resistance is the sum of the individual resistances:

$$R_{eq} = R_1 + R_2 + ... + R_N \qquad (2.1)$$

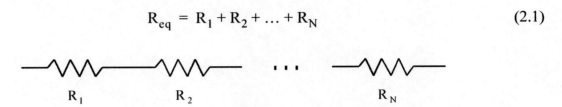

Figure 2.1 Series Resistors

For resistors connected in parallel,

$$\frac{1}{R_{eq}} = \frac{1}{R_1} + \frac{1}{R_2} + ... + \frac{1}{R_N} \qquad (2.2)$$

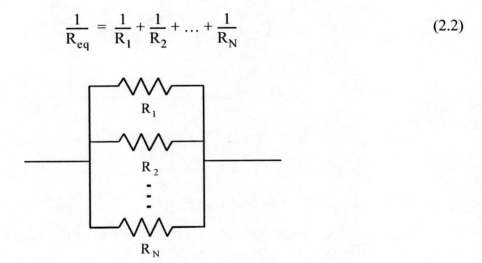

Figure 2.2 Parallel Resistors

For two resistors in parallel, Equation 2.2 can be written as:

$$R_{eq} = \frac{R_1 R_2}{R_1 + R_2} \qquad (2.3)$$

2.2.2   Kirchoff's Voltage Law (KVL)

Kirchoff's Voltage Law (KVL) states that the sum of the voltages around any closed loop must equal zero:

$$\sum_{i=1}^{N} V_i = 0 \tag{2.4}$$

For example, applying KVL (starting at point A) to the circuit shown in Figure 2.3 gives:

$$-V + V_1 + V_2 = 0 \tag{2.5}$$

or

$$V = V_1 + V_2 \tag{2.6}$$

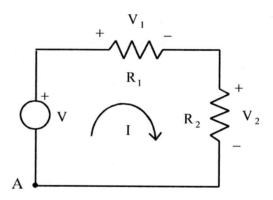

Figure 2.3 Kirchoff's Voltage Law

2.2.3   Kirchoff's Current Law (KCL)

Kirchoff's Current Law (KCL) states that the sum of the currents entering (positive) and leaving (negative) a node must equal zero:

$$\sum_{i=1}^{N} I_i = 0 \tag{2.7}$$

For example, applying KCL to the circuit shown in Figure 2.4 gives:

$$I - I_1 - I_2 = 0 \tag{2.8}$$

18

or

$$I = I_1 + I_2 \qquad\qquad (2.9)$$

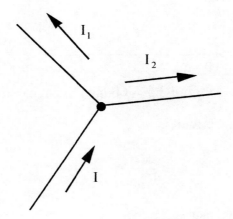

Figure 2.4 Kirchoff's Current Law

2.2.4   Ohm's Law

Ohm's Law states that the voltage across an element is equal to the resistance of the element times the current through it:

$$V = IR \qquad\qquad (2.10)$$

Figure 2.5 Ohm's Law

2.2.5   The Voltage Divider Rule

The voltage divider rule is an extension of Ohm's Law and can be applied to a series resistor circuit shown in Figure 2.6.

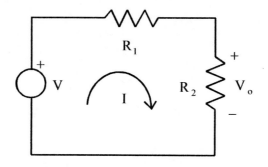

Figure 2.6 Voltage Division

The current flowing in the circuit is

$$I = \frac{V}{R_{eq}} = \frac{V}{R_1 + R_2} \tag{2.11}$$

Applying, Ohm's Law, the voltage across R_2 is

$$V_o = IR_2 \tag{2.12}$$

Thus the voltage divider relation is

$$V_o = V\left(\frac{R_2}{R_1 + R_2}\right) \tag{2.13}$$

2.2.6   The Current Divider Rule

The current divider rule is can be derived by applying Ohm's Law to the parallel resistor circuit shown in Figure 2.7.

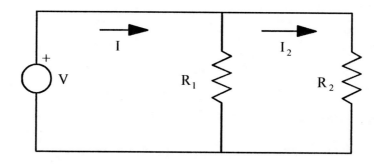

Figure 2.7 Current Division

The current flowing from the voltage supply is:

$$I = \frac{V}{R_{eq}} = \frac{V(R_1 + R_2)}{R_1 R_2} \tag{2.14}$$

Applying Kirchoff's Voltage Law around the outside loop gives:

$$V = I_2 R_2 \tag{2.15}$$

Substituting Equation 2.15 into 2.14 gives:

$$I = \frac{I_2(R_1 + R_2)}{R_1} \tag{2.16}$$

Solving for I_2 gives the current divider relation:

$$I_2 = I\frac{R_1}{R_1 + R_2} \tag{2.17}$$

2.2.7   Root-Mean-Square Values

When dealing with AC signals, voltage and current values can be specified by their root-mean-square (rms) values. An rms value is defined as the square root of the average of the square of a signal integrated over one period. For current and voltage, the rms relations are:

$$I_{rms} = \sqrt{\frac{1}{T}\int_0^T I^2 dt} = \frac{I_m}{\sqrt{2}} \quad \text{and} \quad V_{rms} = \sqrt{\frac{1}{T}\int_0^T V^2 dt} = \frac{V_m}{\sqrt{2}} \tag{2.18}$$

where I_m and V_m are the amplitudes of sinusoidal current and voltage waveforms. Rms values are useful for power calculations. For example, the average AC power dissipated by a resistor can be calculated with the same equations that are used with DC signals:

$$P_{avg} = V_{rms}I_{rms} = RI_{rms}^2 = V_{rms}^2/R \tag{2.19}$$

2.2.8   Real Sources and Meters

When analyzing electrical circuits on paper the concepts of ideal sources and meters are often used. An ideal voltage source has zero output impedance and can supply infinite current. An ideal voltmeter has infinite input impedance and draws no current. An ideal ammeter has zero input impedance and no voltage drop across it. Laboratory sources and meters have terminal characteristics that are somewhat different from the ideal cases. The terminal characteristics of the real sources and meters you will be using in the laboratory may be modeled using ideal sources and meters as illustrated in Figures 2.8 through 2.10

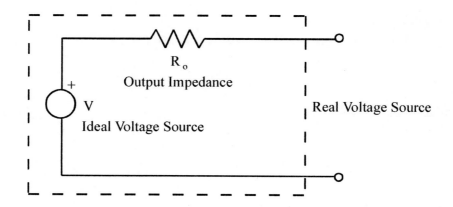

Figure 2.8 Real Voltage Source with Output Impedance

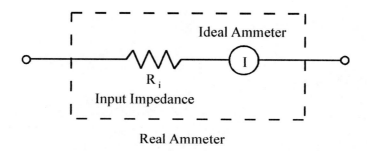

Figure 2.9 Real Ammeter with Input Impedance

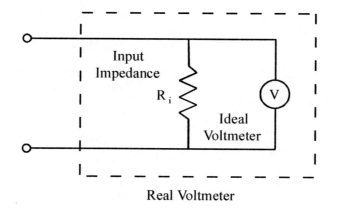

Figure 2.10 Real Voltmeter with Input Impedance

In some instances as you will see, the input impedance of a meter or the output impedance of a source can be neglected and very little error will result. However, in many applications where the impedances of the instruments are of a similar magnitude to those of the circuit serious errors will occur.

As an example of the effect of input impedance, if you use an oscilloscope or multimeter to measure the voltage across R_2 in Figure 2.6, the equivalent circuit is:

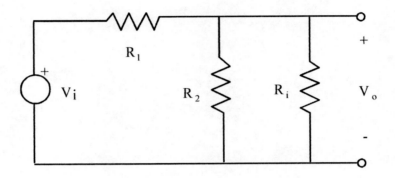

Figure 2.11 Effect of Input Impedance

The equivalent resistance of the parallel combination of R_2 and R_i is:

$$R_{eq} = \frac{R_2 R_i}{R_2 + R_i} \tag{2.20}$$

Therefore, the actual measured voltage would be:

$$V_o = \frac{R_{eq}}{R_1 + R_{eq}} V_i \tag{2.21}$$

If R_i is large compared to R_2 (usually the case), $R_{eq} \approx R_2$ and the measured voltage (V_o) would be close to the expected ideal voltage division result of $\frac{R_2}{R_1 + R_2} V_i$. However, if R_2 is not small compared to R_i, the measured voltage will differ from the ideal result based on Equations 2.20 and 2.21.

If you know values for V_i, R_1, and R_2 in Figure 2.11, and if you measure V_o, you can determine the input impedance (R_i) of the measuring device using the following analysis. Equation 2.21 can be solved for R_{eq} giving:

$$R_{eq} = \left(\frac{V_o}{V_i - V_o}\right) R_1 \tag{2.22}$$

23

Knowing R_{eq}, we can determine the input impedance by solving for R_i in Equation 2.20:

$$R_i = \frac{R_{eq}R_2}{(R_2 - R_{eq})} \qquad (2.23)$$

2.3   Laboratory Procedure / Summary Sheet

Group: _____   Names: _____

(1)   Select five separate resistors whose nominal values are listed below. Read the color code and record the specified value for each resistor in the table below. Then connect each resistor to the multimeter using alligator clips and record the measured value for each resistor.

Resistor	Specified Value (Ω)	Measured Value (Ω)
R_1: 1kΩ	1kΩ	.9860 kΩ
R_2: 1kΩ	1kΩ	.9890 kΩ
R_3: 2kΩ	2kΩ	1.949 kΩ
R_4: 1MΩ	1MΩ	1.012 MΩ
R_5: 1MΩ	1MΩ	1.003 MΩ

(2)   Now construct the voltage divider circuit shown using resistors R_1 and R_2 listed above and set V_i to 10 Vdc using the DC power supply. **When using a power supply or function generator, always adjust the supply voltages before making connections to the circuit.**

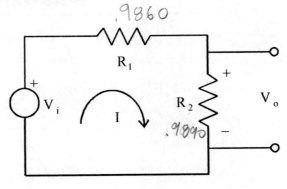

.9860

Figure 2.12 Voltage Divider Circuit

$$V = IR_1 + IR_2$$

$$10V = I(.9860) + I(.9890) \quad \therefore \quad I = 5.06 \, mA$$

25

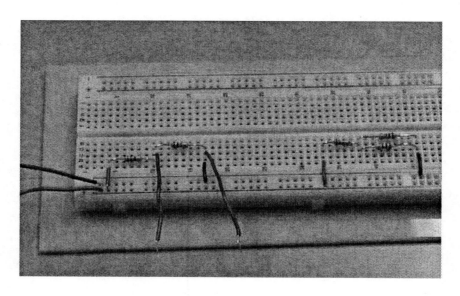

Figure 2.13 Breadboard layout for voltage divider (left) and current divider (right)

Complete the table below by measuring or calculating the appropriate values. In your calculations, use the actual (measured) values for R_1 and R_2.

	Input Voltage V_i (V)	Output Voltage V_o (V)	Current (mA)
Calculated	10 V	5.00	5.06
Multimeter	10.01V	5.030	5.074
Oscilloscope	10V	5V	5.06 *

* compute the current using the voltage value measured

$$V_o = I R_2 = (5.06)(.9890) = 5.00 \checkmark$$

$R_{eq} =$

26

(3) Repeat part 2 using the same resistors R_1 and R_2 but using the function generator to drive the circuit at 1KHz with a 3V amplitude (6V peak-to-peak) sine wave.

Complete the table below by measuring or calculating the appropriate values. In your calculations, use the actual (measured) values for R_1 and R_2. Use rms values for all table entries.

	Input Voltage (V_{rms})	Output Voltage (V_{rms})	Current (I_{rms} in mA)	
Calculated	$\frac{3V}{\sqrt{2}}$	$\frac{1.5}{\sqrt{2}} = 1.06$	1.07	
Multimeter	2.36	1.18	1.19	*
Oscilloscope	$3/\sqrt{2}$	$2/\sqrt{2} = 1.41$	1.07	*

$\dfrac{2.36}{(.9890 + .9860)}$

$\dfrac{3/\sqrt{2}}{(.9890 + .9860)}$

* compute the current using the voltage value measured

$$3v = I(.9860) + I(.9890) \qquad I = \frac{3v}{(.9860 + .9890)} = 1.5 \, mA$$

$$I_{rms} = \frac{1.52}{\sqrt{2}} = 1.07$$

(4) Repeat part 2 ($V_i = 10$ Vdc) using R_4 and R_5 in place of R_1 and R_2. In this case, the impedances of the instruments are close in value to the load resistances and therefore affect the measured values. Sketch the equivalent circuit for the instruments and the attached load circuit. Use this schematic to explain differences between actual and measured values.

$$\frac{1}{R_e} = \frac{1}{R_5} + \frac{1}{R_4} = \frac{2}{1M\Omega} \qquad R_e = .5 M\Omega$$

$V = IR$

$I = \frac{V}{R}$

Complete the table below by measuring or calculating the appropriate values. In your calculations, use the actual (measured) values for R_4 and R_5.

	Input Voltage (V)	Output Voltage (V)	Current (mA)	
Calculated	10	2.006	.002	
Multimeter	10.03	4.761	.0063 mA	
Oscilloscope	10	3.5	.010	*

* compute the current using the voltage value measured

$$10v = I(1M\Omega) + I(1M\Omega) \qquad I = .002 \, mA \qquad V_o = R_5 I$$

$$= (1.003 M\Omega)(.002 \, mA)$$

$$= 2.006 \, V$$

$R_{eq} = 1.015 M\Omega$

(5) Construct the current divider circuit shown below using resistors R_1, R_2, and R_3 listed in part 1. Set the source V to 6 Vdc.

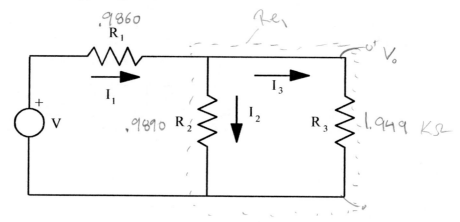

Figure 2.14 Current Divider Circuit

Complete the table below by measuring or calculating the appropriate values. In your calculations, use the actual (measured) values for R_1, R_2, and R_3.

	I_1 (mA)	I_2 (mA)	I_3 (mA)
Calculated	3.654	2.424	1.23
Multimeter	3.648	2.416	1.224
Oscilloscope	3.04 *	2.02 *	1.03 *
	V = 3	V = 2	V = 2

* compute the current using the voltage value measured

(6) Repeat part 5 with a 3 V amplitude 500 Hz sine wave ($V = 3\sin(1000\pi t)$).

Complete the table below by measuring or calculating the appropriate values. In your calculations, use the actual (measured) values for R_1, R_2, and R_3. Use rms values for all table entries.

	I_{1rms} (mA)	I_{2rms} (mA)	I_{3rms} (mA)
Calculated	1.292	.857	.4349
Multimeter	1.63 *	.7795 *	.3956 *
Oscilloscope	1.43 *	.715 *	.363 *
	$V_R = 1.61$ $V_o = 2$	$V_m = .771$ $V_o = 1$	

* compute the current using the voltage value measured

As has been demonstrated, the input impedance of a meter or the output impedance of a source can be neglected and very little error will result. However, in some applications where the impedances of the instruments are of a similar magnitude to those of the circuit serious errors will occur.

$V = IR$

$I = \frac{V}{R}$

$R_{e_1} = \frac{R_2 R_3}{R_2 + R_3} = \frac{(.9890)(1.949)}{.9890 + 1.949} = .6561$

$R_{eq} = .6561 + .9860 = 1.642 \ k\Omega$

$6V = I_1(1.642) \qquad I_1 = 3.654 \ mA$

$I_3 = I_1\left(\frac{R_2}{R_2 + R_3}\right) = 3.654\left(\frac{.9890}{.9890 + 1.949}\right) = 1.23$

28   $I_2 = 3.654\left(\frac{1.949}{1.949 + .9890}\right) = 2.424$

LAB 2 QUESTIONS

Group: _____   Names: _____

(1)   Describe how you read resistor values and tolerances.

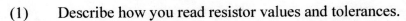

(2)   Write down the voltage divider rule and the current divider rule.

V = _____

I = _____

(3)   From the data collected in Part 4, calculate the input impedance of the oscilloscope and the DMM.

Z_{in} (scope) = _____

Z_{in} (DMM) = _____

Hint: use Equations 2.22 and 2.23.

(4)   The AC wall outlet provides 110 V_{rms} at 60Hz. Sketch and label one period of this waveform.

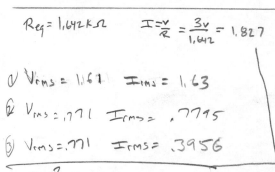

$R_{eq} = 1.642 k\Omega$ $I = \frac{V}{R} = \frac{3v}{1.642} = 1.827$ $I_{rms} = \frac{1.827}{\sqrt{2}} = 1.292$

① $V_{rms} = 1.61$ $I_{rms} = 1.63$

② $V_{rms} = .771$ $I_{rms} = .7715$

③ $V_{rms} = .771$ $I_{rms} = .3956$

$I_1 = \frac{2}{.9860} = 2.02$ $I_{rms} = 1.43$

$I_2 = \frac{1}{.9890} = 1.01$ $I_{2rms} = .715$

$I_2 = 1.827\left(\frac{1.949}{1.949 + .9890}\right) = 1.21$

$I_{2rms} = \frac{1.21}{\sqrt{2}}$

$I_3 = 1.827\left(\frac{.9890}{1.949 + .9890}\right)$ $= .857$

$= .615 mA$

$I_{rms} = \frac{.615}{\sqrt{2}} = .4349$

$I_3 = \frac{1}{1.949} = .5131$ $I_{3rms} = .363$

(5) Using a function generator and three 1 kΩ resistors design a circuit that will supply both a 6V p-p output and a 2V p-p output. Show your work below.

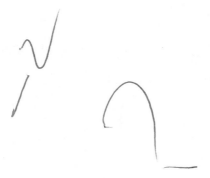

Laboratory 3

The Oscilloscope

Required Components:
- 1 10Ω resistor
- 2 100Ω resistors
- 2 1kΩ resistors
- 1 2kΩ resistor
- 2 4.7MΩ resistors
- 1 0.022μF capacitor
- 1 0.1μF capacitor
- 1 1.0μF capacitor

3.1   Objectives

In the previous laboratory exercise you learned about the basic operation of the oscilloscope. This laboratory exercise is designed to give you a more in-depth understanding of the proper use of the oscilloscope and its range of applications.

The oscilloscope is probably one of the most widely used electrical instruments and is one of the most misunderstood. During the course of this laboratory exercise you will become familiar with the proper methods of connecting inputs, grounding, coupling, and triggering the oscilloscope. Also during the course of this experiment you will learn the proper use of the oscilloscope attenuator probe.

3.2   Introduction

3.2.1   AC and DC Signals

An AC signal varies with time, and its deterministic expression contains time as the independent variable. For example,

$$F_1(t) = 2.0 \sin 5t \tag{3.1}$$

$$F_2(t) = 3.1 \cos 5t + 5.1 e^{-3.0t} \tag{3.2}$$

A DC signal on the other hand does not vary with time, hence t does not appear in its expression:

$$F_3(t) = 1.0 \tag{3.3}$$

$$F_4(t) = 5.63 \tag{3.4}$$

Now what if our signal can be written:

$$F_5(t) = 2.0 + 1.0 \sin 5t \qquad (3.5)$$

Is it AC or DC? Well, we say it is AC (1.0 sin 5 t) with a DC offset (2.0). We can see this if we plot the signal below:

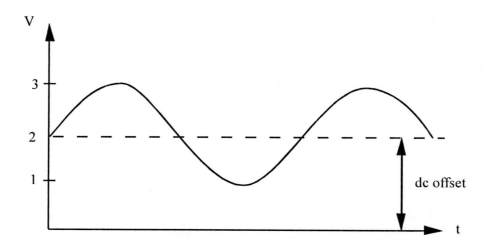

Figure 3.1 AC Signal with DC Offset

This difference between an AC and DC signal is important when understanding oscilloscope coupling.

3.2.2   AC and DC Coupling

Most oscilloscopes are provided with a switch to select between AC or DC coupling of a signal to the oscilloscope input amplifier. When AC coupling is selected, the DC component of the signal is blocked by a capacitor inside the oscilloscope that is connected between the input terminal and the amplifier stage. Both AC and DC coupling configurations are illustrated in Figure 3.2. R_{in} is the input resistance and C_{in} is the input capacitance. C_c is the coupling capacitor that is present only when AC coupling is selected.

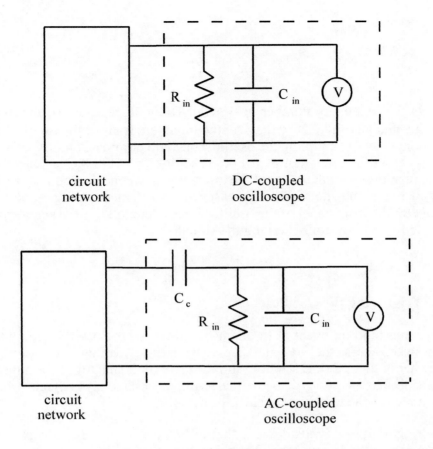

circuit
network

DC-coupled
oscilloscope

circuit
network

AC-coupled
oscilloscope

Figure 3.2 Oscilloscope Coupling

<u>AC Coupling</u>

AC coupling must be selected when the intent is to block any DC component of a signal. This is important, for example, when measuring small AC spikes and transients on a 5 V TTL (transistor-transistor logic) supply voltage. However, it must be kept in mind that with AC coupling:

- One is not aware of the presence of any DC level with respect to ground.

- The lower frequency components of a signal are attenuated.

- When the oscilloscope is switched from DC to AC coupling, it takes a little time before the display stabilizes. This is due to the time required to charge the coupling capacitor C_c to the value of the DC component (average value) of the signal.

- Sometimes the input time constant ($\tau = R_{in}C_c$) is quoted among the oscilloscope specifications. This number is useful, because after about five time constants (5τ), the displayed signal is stable.

AC coupling can be explained by considering the impedance of the coupling capacitor as a function of frequency:

$$Z = \frac{1}{j\omega C} \qquad (3.6)$$

where j represents the imaginary number $\sqrt{-1}$. For DC voltages ($\omega = 0$) the impedance of the capacitor is infinite, and all of the DC voltage at the input terminals of the oscilloscope will appear across the capacitor. Thus AC coupling the oscilloscope will eliminate any DC offset present in the voltage appearing across the input terminals of the oscilloscope. For AC signals, the impedance is less than infinite, resulting in attenuation of the input signal dependant upon the frequency. As the input frequency increases the attenuation decreases to zero. The coupling mode is selected using the input selectors on the front panel of the oscilloscope. Generally, if the signal type is unknown, DC coupling is the first choice for observing the signal.

3.2.3   Triggering the Oscilloscope

Triggering refers to an event at the input terminals of the oscilloscope which causes the electron beam to sweep across the CRT and display the terminal voltage. The oscilloscope may be level triggered either in the AC or DC mode, and the level of the magnitude is adjustable using the trigger level control. The slope (+ or −) of the terminal voltage also affects when the beam is triggered. This slope is selected either positive or negative.

Another triggering option available is that of line triggering. Line triggering uses the AC power input to synchronize the sweep. Thus any terminal voltage synchronized with the line frequency of 60Hz or multiples of 60Hz can be triggered in this mode. This is useful because we can detect if 60 Hz noise from various line related sources is superimposed on the signal.

3.2.4   Grounding Source and Scope

Normally, all measurement instruments, power sources, and signal sources in a circuit must be referenced to a common ground as shown in Figure 3.3.

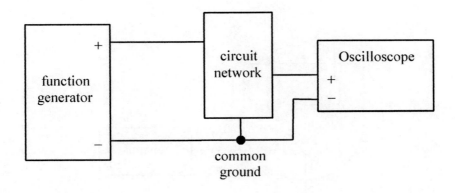

Figure 3.3 Common Ground Connection

However, as can be seen from Figure 3.4, if we wish to measure a differential voltage ΔV, it is correct to connect the scope as shown. Note that the oscilloscope signal ground and external network ground are not common. This type of connection allows us to measure a potential difference anywhere in a circuit.

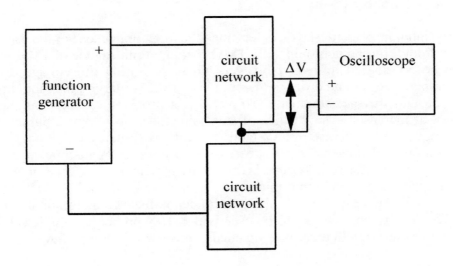

Figure 3.4 Relative Ground Connection

3.2.5 Properly Grounding the Oscilloscope to the Wall Socket

As with most of the instruments you will be using in the laboratory the oscilloscope is equipped with a 3-pronged plug (see Figure 3.5) for safety purposes. The two flat prongs of this plug complete the circuit for alternating current to flow from the wall socket to the instrument. The round prong of the plug is connected only to the chassis and not to the signal ground. This is important to protect the operator if there is a short circuit inside the oscilloscope. Otherwise, a high voltage can occur on the chassis jeopardizing the safety of the user.

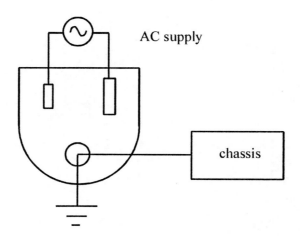

AC supply

chassis

Figure 3.5 Three-prong AC Power Plug

3.3   Using the Attenuator Probes

In the first laboratory exercise you determined the input impedance of the oscilloscope, and you should have found it to be approximately 1MΩ. An input impedance of 1MΩ is large and in most cases can be considered infinite. However, when measuring the voltage drop across an element whose impedance is of an order of magnitude of 1 MΩ or larger, the input impedance can induce serious error in the measurement. To avoid this problem, the input impedance of the oscilloscope must be increased. One method of increasing the oscilloscope input impedance is the use of an attenuator probe. The use of an attenuator probe will increase the input impedance by some known factor but will at the same time decrease the amplitude of the input signal by the same factor since the current into the oscilloscope is limited by the input impedance. Thus a 10X probe will increase the magnitude of the input impedance of the oscilloscope by a factor of 10, but the displayed voltage will be only 1/10 of the amplitude of the actual terminal voltage. Most oscilloscopes offer an alternative scale to be used with a 10X probe. A simple schematic of the oscilloscope input terminals with the probe attached is presented in Figure 3.6.

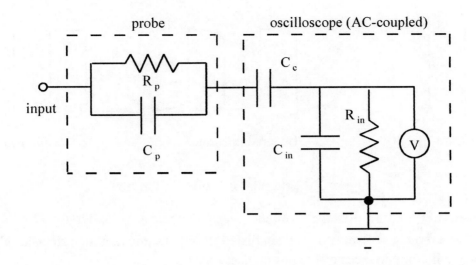

Figure 3.6 Simple Model of Oscilloscope and Probe

Note that the addition of the attenuator probe to the input terminals of the oscilloscope not only changes the resistive characteristics of the terminals but the capacitive characteristics as well. A complete model for the oscilloscope, the cable connections and the attenuator probe is shown in Figure 3.7.

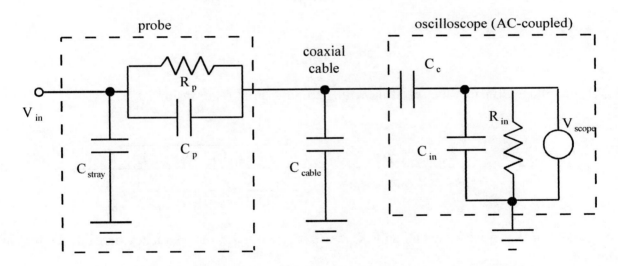

Figure 3.7 Complete Model of Oscilloscope, Probe, and Cable

Due to the collection of complex impedances between the input (V_{in}) and the oscilloscope voltage measuring device (V_{scope}), the voltage reading will depend on the frequency components of the input (in addition to the input voltage magnitude). However, by adjusting C_p, this dependence can be minimized. C_p can be adjusted by turning the small screw in the attenuator probe and monitoring a square wave output from the probe adjust port on the front panel of the oscilloscope. The diagram given in Figure 3.8 will help you when tuning the probe.

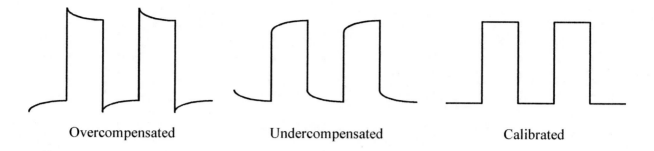

Overcompensated Undercompensated Calibrated

Figure 3.8 Attenuator Probe Calibration

The use of C_p to minimize the frequency dependence can be understood by looking at the impedances in the model. To simplify the analysis we will assume that the effects of C_{stray} and C_c are negligible, so the impedances of the circuit sections are

$$Z_{scope} = \frac{R_{in}}{j\omega R_{in}C_{in} + 1} \qquad (3.7)$$

$$Z_{scope + cable} = \frac{R_{in}}{j\omega R_{in}(C_{in} + C_{cable}) + 1} \qquad (3.8)$$

$$Z_{probe} = \frac{R_p}{j\omega R_p C_p + 1} \qquad (3.9)$$

and from voltage division, the resulting voltage input-output relationship is

$$\frac{V_{scope}}{V_{in}} = \frac{Z_{scope + cable}}{Z_{probe} + Z_{scope + cable}} = \frac{\dfrac{R_{in}}{j\omega R_{in}(C_{in} + C_{cable}) + 1}}{\dfrac{R_p}{j\omega R_p C_p + 1} + \dfrac{R_{in}}{j\omega R_{in}(C_{in} + C_{cable}) + 1}} \qquad (3.10)$$

Now it is clear that if we adjust C_p so $R_p C_p = R_{in}(C_{in} + C_{cable})$, the frequency dependence will be eliminated.

3.4   Specific Background Information on Using an HP54602A Oscilloscope

Looking at the front of the scope you can see that it contains a screen and a number of buttons and knobs. The screen displays the output of the oscilloscope. Just below the screen are six 'soft' buttons. The function of these buttons will change depending upon menu selections. The current function of each button is displayed directly above the button on the bottom of the screen.

3.4.1 The Screen

The screen is divided into eight vertical and ten horizontal divisions. Typical information displayed on the screen includes:

The Y-Axis Scale: The Y-Axis scale is located in the upper left hand corner of the screen. Each channel can have a different scale. For example if the screen reads: 1 2V 2 200mV, it means that channel one has a scale of 2 volts per division, while channel two has a scale of 200 millivolts per division.

The Time Scale: The time scale is located in the upper right side of the screen. Each channel has the same time scale. For example, 10ms/ means 10 milliseconds per division.

Delay: The time delay indicator is located in the upper middle section of the screen. This displays whether the signal is shifted left or right. Typically this should be reading 0.00 s.

Trigger information: The current trigger mode is in the upper right hand corner of the screen. It is displayed either as an up arrow or as a down arrow indicating positive or negative slope triggering.

Channel Grounds: Along the right hand side of the display there will be a channel ground indicator (a ground symbol with the channel number next to it). It shows where the ground for that channel is with respect to the displayed grid. If the ground is off the screen there will be an arrow with the channel number at the top or bottom of the screen to indicate where the ground is.

Along the bottom of the screen there is room for addition information to be displayed. The very bottom of the screen indicates the current functions of the soft buttons. These functions change depending upon which menu button has been selected.

3.4.2 Buttons and Dials

The buttons and dials on the oscilloscope are separated into five groups: Vertical (outlined in blue), Horizontal (outlined in gray), Trigger (outlined in green), Storage (outlined in gray) and an unnamed group (also outlined in gray). The unnamed group contains the Measure and Save/ Recall sub-groups as well as the Autoscale, Display, and Print/Utility buttons.

3.4.3 Vertical

The vertical group contains four columns, one for each channel. Each column contains a coaxial input port. Above the input port is a dial labeled "Level." This dial can be used to adjust the location of the ground on the screen. Above the "Level" knob there is a button with the channel number on it. Pressing this button will call up a menu that can be accessed with the soft buttons located just below the screen. For channels 1 and 2 this menu includes:

Off On: Selecting this will toggle the trace of the selected channel on and off

Coupling: Allows switching between AC and DC coupling.

BW Limit: Allows limiting the bandwidth of the signal displayed (helps to reduce noise).

Vernier: Leave off.

Probe: This indicates to the oscilloscope if an attenuation probe is attached. The scope will multiply the incoming signal by the selected value before displaying it.

Channels 3 and 4 have less extensive capabilities:

Off On: Same function as channels 1 and 2

Coupling: Only DC coupling is available for channels 3 and 4

V/div: Channels three and four are limited to 2 possible voltage scales

Probe: Same as channels one and two

Between the channel 1 and channel 2 buttons, there is a button with a plus and a minus on it. This button will bring up a menu that allows you to perform math on the signals. You can add, subtract and multiply signals together or perform and display the results (spectrum) of a Fast Fourier Transform (FFT).

For channels 1 and 2 there is also a large green knob. This knob will change the Y-axis scale on the screen.

3.4.4   Horizontal

The horizontal control group contains one large knob (Time/Div), one small knob (Delay), and one button labeled Main/Delay. The large knob is used to adjust the time scale on the screen. This will allow you to zoom in on sections of the signal. The small knob labeled "Delay" allows you to move the signal left and right so you can inspect different areas of the signal. The button brings up a menu that allows you to choose one of four horizontal modes and switch the location of your time reference from the center of the screen to the left of the screen.

The four horizontal modes are:

Main: This is the normal or default setting you will usually use.

Delayed: This splits the screen vertically into two windows. On the top you will see a large section of your signal separated by two vertical lines. On the bottom window you will see an expanded version of the signal that lies within the two vertical lines on the upper window. By adjusting the delay knob you can move left and right through the signal. Adjusting the Time/Div changes the spacing between the two vertical lines and thus the amount of "zoom."

XY: This allows you to use channel 1 as the x-axis instead of time.

Roll: In this mode the scope just displays the current signal without 'trying' to maintain a steady display.

3.4.5   Trigger

Triggering ensures that the display of a signal will be stable for direct observation.

Triggering is the event that causes the scope to start scanning the electron beam creating the displayed signal called a trace. The scope starts the trace at the left and moves to the right at a speed determined by the time scale. For example, if the time scale is 200ms/division it will take the scope 2 ms (200ms/division times ten divisions) to complete the trace. Once the trace has reached the end of the screen the scope waits for another triggering event to initiate the trace again. The triggering event can be adjusted by the knobs and buttons in the trigger menu.

The trigger group contains two knobs and three buttons:

Source: This button allows you to choose which channel will be used as the trigger. In addition to the four input channels, the line can also be used as a trigger. This means the scope will trigger based on the voltage coming from the AC power source. This is useful when looking for noise caused by electrical interference.

Mode: You should usually select the normal mode

Slope/Coupling: This button will bring up a menu that allows you to select the slope of the trigger. You can either trigger on the signal when it is going up or going down (up arrow, down arrow) depending on how you want the signal displayed. This menu also contains features that will help you trigger on a noisy signal

Level: This knob allows you to adjust the voltage level at which triggering occurs. Combined with the slope this is how triggering is adjusted. For example, if you have the level set to zero and the slope set to + (up arrow) you will be triggering every time the signal goes from below zero volts to above zero volts. The level is very important when you are dealing with signals that have been rectified (by a diode). A rectified signal may never have a value of less than zero, and if you leave the level at zero, the scope may never trigger, and no signal will be displayed.

Holdoff: You should not need to use this.

3.4.6   Storage

The scope has the ability to store signals for later display and comparison.

3.4.7   Measure

There are three buttons and some knobs in the measure sub-group:

Voltage: pressing this button brings up a menu that will allow you to make a variety of voltage measurements.

By selecting the appropriate soft button you can display just about any information that you want. For example, if you want to know the RMS voltage of the signal on channel 2, you should push the left-most soft button until the number 2 is highlighted and then press the Vrms button. The RMS voltage will appear near the bottom of the screen.

Time: Similar to the voltage button, this displays a menu that will allow you to make a variety of time measurements. You can measure frequency, period, rise time, etc.

Cursors: This button allows you to make manual measurements of a signal. It will bring up a window with the following options.

Source: Allows you to change the channel you are measuring.

Active Cursor: You will be able to have four cursors on the screen at once. Two are vertical labeled V1 and V2, and two are horizontal labeled T1 and T2. Using the unlabeled knob just below and to the right of the cursors button you will be able to move the active cursor around. While you are moving the cursor the screen will display its position relative to ground (or zero time) and also relative to the other cursor. This allows you to measure the voltage difference for any specific part of the signal.

3.4.8   Save/Recall

This group contains two buttons:

Trace: This allows you to save a trace so you can look at it later.

Setup: This button brings up a menu that allows you to save the setup of your voltage and time scales as well as your triggering. This may be useful if you have a particular setup you like. Also in this menu is an Undo Autoscaling button that will be useful if you press the Autoscale button by accident and lose your signal.

Autoscale

This is probably the most useful button on the scope. Pressing this button will automatically setup the voltage scales, time scale, and triggering so you produce a stable display. Although it is very helpful, it is not perfect for every application. It is easy to overuse this button. Here are some things to watch out for:

(1)   If you have a relatively low frequency signal (less than 50 Hz) Autoscale will not find the signal.

(2)   If you have a DC signal, Autoscale may not find the signal. If so, try switching your trigger source to line if you cannot get a DC signal displayed.

(3)   If you are using more than one channel Autoscale will set the vertical scales differently and adjust the ground levels so each signal is displayed separately.

Display

This button will bring up a menu with the following options:

Display Mode [normal, peak detect, average]: Leave this on normal

Vernier: Leave this off

Grid: Pressing this will let you turn the grid on and off

3.4.9   How to Find a Signal on an HP54602A Oscilloscope

Use the following procedure to display an unknown input signal on the oscilloscope:

(1)   First try pressing the Autoscale button. Often this will automatically scale and display the signal. If the signal is not displayed (e.g., when a signal has low frequency), continue with the remaining steps below.

(2)   Make sure the desired channel is on and set up properly.

- Press the number button corresponding to the channel you want to observe

- Turn the channel on by pressing the left-most "soft" button.

- With the right-most 'soft' button, select the type of probe you are using.

- If it is possible that your signal could have a large DC offset, select AC coupling with the second to left "soft" button.

- Make sure BW Lim, Invert, and Vernier are off.

(3)   Move the channel ground to the center of the display.

- There should be a ground symbol on the right hand side of the screen.

- If the ground is off the screen there will be an arrow in either the upper or lower right hand corner pointing to it.

(4)   You can move the ground up and down with the small, light colored, "position" knob. Make sure the trigger is set up correctly.

- Press the 'source' button and then select the appropriate channel

- Press the 'mode' button and select "Auto Lvl"

- Adjust the level knob until the level is within the range of the signal

- If you have a noisy signal press the Slope/Coupling button and turn the "Noise Rej" on.

(5)   Set the Vertical scale.

- The vertical scale is displayed in the upper left corner of the screen

- Turn the Volts/Div knob (above Channel 1) until the vertical scale is about half of the amplitude of the signal.

(6)   Set the Time scale.

- The time scale is displayed at the top of the screen (just right of center)

- Turn the Time/Div knob until the time scale is about ½ of the period of the signal.

3.5   Laboratory Procedure / Summary Sheet

Group: _____   Names: _____

(1)   In this step you will examine the effects of AC coupling of an oscilloscope.

Adjust the function generator to produce a 5 V_{pp} 1kHz sinusoidal output.

Display the sinusoid on the oscilloscope with a sensitivity of 1 V/cm. Adjust the triggering level and note what you see.

Use the +/– trigger selector (press the slope button and toggle between positive and negative edge triggering) and note what happens.

Trigger the oscilloscope using the channel and line triggering inputs and again note what you see, in each case.

Refer back to Figure 3.2 for details on the impedances within the oscilloscope. We can simulate an AC coupled oscilloscope by adding an external coupling capacitor to a DC coupled oscilloscope. To do this, build the circuit shown in Figure 3.9. Add a 1 Vdc offset to the 5V_{pp} 1kHz signal in the output of the function generator. Be sure to select DC coupling on the oscilloscope. Use a 0.022 μF coupling capacitor ($C_{coupling}$) and note the resulting output. Then try a 0.1 μF and a 1 μF capacitor. Note carefully what happens to the oscilloscope display when you first attach the scope probe to measure V_o for each case. Make sure you discharge the coupling capacitor by shorting its leads before you attach the probe to measure the voltage. How do the results change with the capacitance value? What is the effect of the function generator output impedance R_o on the measured voltage V_o?

Now remove the coupling capacitor and toggle the oscilloscope between DC coupling and AC coupling and note what you observe in the measured voltage signal. Which of the three external coupling capacitors that you tried most closely approximates what you think the actual coupling capacitance of the oscilloscope is?

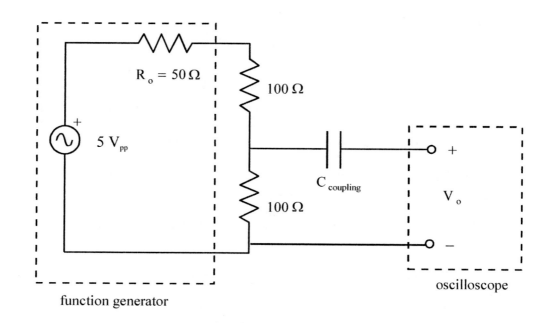

Figure 3.9 Coupling Capacitor Effects

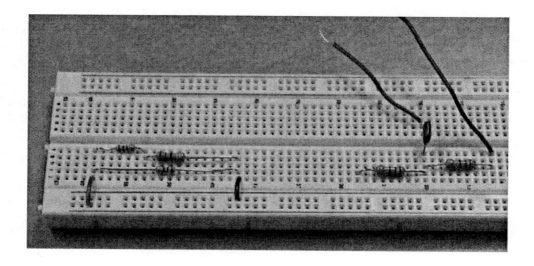

Figure 3.10 Circuits for Steps 1 (right) and 4 (left)

(2) In this step and the next you will study the effects of the input impedance of the oscilloscope. Also, you will use an oscilloscope probe to increase the input impedance and note the consequences.

Construct the circuit shown below with $R_1 = R_2 = 1 \text{ k}\Omega$ and $V_i = 6V_{pp}$ at 1kHz.

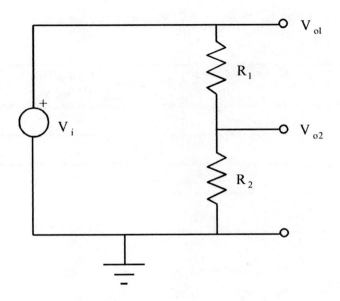

Figure 3.11 Probe Measurements

Connect a 10X attenuator probe to channel 1 and calibrate the probe. Use the probe adjust port on the oscilloscope.

Using the attenuator probe determine each of the voltages V_{o1} and V_{o2}, for the circuit above and record the values in the following table. Be sure that nothing is connected to Channel 2.

	Calculated	Measured with 1X probe	% Error	Measured with 10X probe	% Error
V_{o1}					
V_{o2}					

$R_1 =$ _____

$R_2 =$ _____

(3) Repeat Part II for $R_1 = R_2 = 4.7$ MΩ and record the values V_{o1} and V_{o2} in the following table.

	Calculated	Measured with 1X probe	% Error	Measured with 10X probe	% Error
V_{o1}					
V_{o2}					

$R_1 = $ _____

$R_2 = $ _____

(4) Although the oscilloscope is primarily a voltage measuring instrument it can be used to indirectly measure current by inserting a small value resistor in the circuit branch of interest (unless there is a resistor in the branch already). In order to measure the current we use the oscilloscope to measure the voltage drop across this resistor and then the current through it can be calculated using Ohm's Law.

For the network shown below determine the current (I) by inserting the resistor (R) and measuring the voltage drop across it.

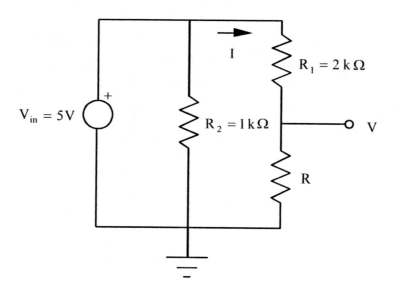

Figure 3.12 Measuring Current

Using the values of R shown in the table below, measure the voltage (V) and from the voltage calculate the current (I) using Ohm's Law.

R	Calculated I*	Measured with 1X probe	% Error	Measured with 10X probe	% Error*
0 Ω					
10 Ω					
100 Ω					
1 KΩ					

* based on R=0 Ω

$R_1 =$ _____

$R_2 =$ _____

LAB 3 QUESTIONS

Group: _____   Names: _____

(1)   From the data obtained in this laboratory exercise determine the input impedance of the oscilloscope with the attenuator probe attached.

(2)   How does the input impedance of the oscilloscope with the 10X attenuator probe compare to the input impedance with the 1X probe?

(3)   By what factor is the input voltage attenuated when the 10X probe is used with the oscilloscope?

(4)   When is it advantageous to use the oscilloscope attenuator probe?

(5)   You have a 4 V$_{pp}$, 100 Hz sine wave, and I wish it to be displayed on the oscilloscope as shown below. Describe how to set the vertical amplifier, time base and trigger.

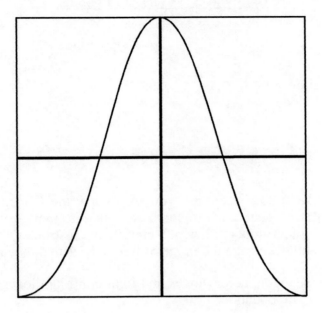

Figure 3.13 Oscilloscope Sine Wave Display

(6)   What is the effect of AC coupling when a signal with a DC offset is displayed on an oscilloscope?

(7)   In part 4, what effect does the inserted resistor have on the current measured?

Laboratory 4

Bandwidth, Filters, and Diodes

Required Components:
- 1 1kΩ resistor
- 1 0.1μF capacitor
- 1 small signal diode
- 1 LED

4.1   Objectives

In the previous laboratory exercise you examined the effects of input and output impedances of instruments on signals and measurements. In this laboratory exercise you will study the bandwidth, which is another important signal, circuit, and instrument characteristic. You will build basic filter circuits and determine the range of frequencies that they affect.

You will also use semiconductor diodes and light emitting diodes (LED) and build basic circuits that require these components.

4.2   Introduction

Ideally an instrument with purely resistive input terminal characteristics should be able to faithfully reproduce any input of any frequency. However, real instruments also have capacitance and inductance which affect the quality of signal reproduction. With real instruments the range of frequencies over which the input is faithfully reproduced is limited, quite often severely, by such factors as reactance (capacitance or inductance) in electrical systems, and inertia and damping in mechanical systems.

In order to quantify the range of frequencies a system can reproduce, the term bandwidth is used. The bandwidth of a system is defined as: the range of frequencies for which the amplitude of the input of the system is attenuated not more than 3 dB. This is equivalent to 70.7% of its original value. The frequency at which the gain of the system drops below −3 dB is defined as a corner or cutoff frequency ω_c.

The bandwidth of a deterministic electrical system can be readily determined analytically by writing the transfer function and solving for the corner frequency as is done below for the for the example circuit shown in Figure 4.1.

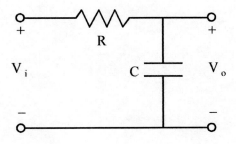

Figure 4.1 Low Pass Filter

Using the voltage divider relationship for AC circuit analysis, where the complex impedance of the resistor is R and the complex impedance of the capacitor is $\frac{1}{j\omega C}$, the output voltage for the network can be written in terms of frequency as

$$V_o = V_i \frac{\frac{1}{j\omega C}}{\frac{1}{j\omega C} + R} \qquad (4.1)$$

and the transfer function (complex output amplitude divided by input amplitude) is

$$T(j\omega) = \frac{V_o}{V_i}(j\omega) = \frac{1}{j\omega RC + 1} \qquad (4.2)$$

The amplitude ratio (V_o/V_i) is the magnitude of the transfer function:

$$\frac{V_o}{V_i}(\omega) = |T(j\omega)| = \frac{1}{\sqrt{1 + (RC\omega)^2}} = \frac{1}{\sqrt{1 + \left(\frac{\omega}{\omega_c}\right)^2}} \qquad (4.3)$$

which is a real function of frequency ω where

$$\omega_c = \frac{1}{RC} \qquad (4.4)$$

Note the following:

$$\text{as } \omega \rightarrow 0, \quad \frac{V_o}{V_i} \rightarrow 1 \tag{4.5}$$

$$\text{as } \omega \rightarrow \infty, \quad \frac{V_o}{V_i} \rightarrow 0 \tag{4.6}$$

$$\text{as } \omega \rightarrow \omega_c, \quad \frac{V_o}{V_i} \rightarrow \frac{1}{\sqrt{2}} = 0.707 = -3 \text{ dB} \tag{4.7}$$

Therefore, ω_c is the corner or cutoff frequency. This frequency, in Hertz, is

$$f_c = \frac{1}{2\pi RC} \tag{4.8}$$

The bandwidth of this circuit is:

$$0 \leq \omega \leq \omega_c \tag{4.9}$$

which implies that this circuit passes low frequencies only which is why it is called a low-pass filter.

In order to experimentally determine the bandwidth of a circuit it is necessary to drive the circuit with signals having a range of frequencies. Measuring the output over the input as a function of frequency determines the frequency response of the system. The bandwidth is found by finding the −3 dB points of the frequency response curve. If there are two cutoff points, the bandwidth is written as:

$$\omega_{c_{low}} \leq \omega \leq \omega_{c_{high}} \tag{4.10}$$

To determine the bandwidth of the circuits used in this laboratory exercise you will be using a function generator. The function generator has a unique feature which readily lends itself to determining the bandwidth. This is the frequency sweep feature which slowly varies the output frequency from a low frequency value (start frequency) to a high frequency value (stop frequency). Using this feature, a frequency response display can be created directly on an oscilloscope. The frequency response curve is a plot of the output amplitude divided by the input amplitude vs. frequency.

4.3 Using the Frequency Sweep Feature on the PM5193 Programmable Function Generator

This section outlines the procedure to generate a frequency sweep with the PM5193 programmable function generator. A swept frequency is a signal of a specific waveform (e.g., a sine or square wave) whose frequency increases in a step-wise fashion from a selected start frequency to another selected stop frequency as shown in Figure 4.2. The frequency is controlled by a voltage-controlled oscillator in the instrument. The voltage that corresponds to the frequency is available at the SWEEP output on the back of the instrument.

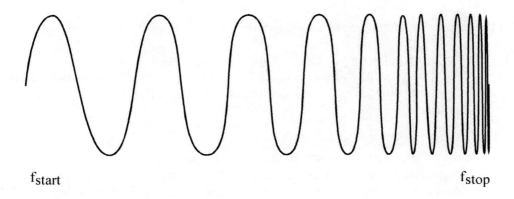

f_{start} f_{stop}

Figure 4.2 Frequency Sweep

The following steps outline the procedure for creating a swept output (in this case, a 10 ms sweep from 1 Hz to 4000 Hz):

(1) <u>Choose the type of waveform you desire</u>
Push the "sine wave" button on the keypad labeled "Wave Form."

(2) <u>Select the "start" frequency</u>
Press the "start" button on the keypad labeled "Frequency." Then type in the numerical value of 1. If Hz is not indicated, press the Hz/kHz button to indicate Hz. **Do not press ENTER until step 7!**

(3) <u>Select the "stop" frequency</u>
Press the "stop" button to the right of the "start" button. Now type in the value of 4000, again making sure that Hz is indicated.

(4) <u>Select the amplitude or peak-to-peak voltage</u>
Press the "V_{pp}" button on the "Level" keypad. Then enter a value of 2 to result in an amplitude of 1.

(5) <u>Set the DC offset</u>
Press the "V_{dc}" button and enter 0.

(6) <u>Select the sweep time</u>
Press the "Time(s)" button under "Modulation" and enter 10 milliseconds (0.01 s).

(7) <u>Activate the parameters</u>
Press the orange "enter" button at the far right of the function generator.

(8) <u>Activate the line sweep</u>
Press the "lin sweep" button on the "Modulation" keypad. Within 4 seconds of pressing the button, enter 1 on the keypad to select the appropriate mode. Then press

the "cont" button to continuously repeat the sweep defined by the parameters above. In order to change any of the parameters, press the "single" button, change the parameter as indicated above, and press the "cont" button to resume the continuously repeating sweep.

Use the following steps to generate a frequency response display on the oscilloscope:

(1) Set the time and voltage scales to appropriate values
Set the time base of the scope to match the sweep time, in this case 2 ms/div. Then select an amplitude/division value (e.g., 0.5 V/div) that will fit the entire waveform on the display.

(2) Use the sweep signal as the trigger
Connect the SWEEP output on the back of the PM5193 to CH2 of the oscilloscope with its vertical scale set at 0.5 V/div. This is a stepped triangular waveform that controls the frequency sweep (see Figure 4.3). Select Norm Trigger, CH2, -slope, and internal source to trigger off the falling edge of the triangle waveform (see the figure below). To better view the full frequency response, place the first step of the triangle on the 2nd division on the scope and the last step on the 6th division (see Figure 4.3). With the parameters defined above, 5 horizontal divisions (2 ms/div) correspond to the sweep time of the signal.

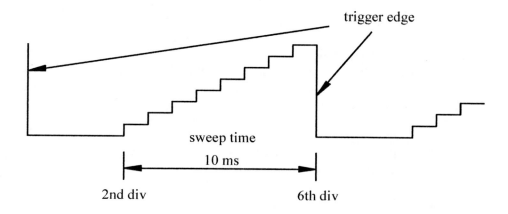

Figure 4.3 Function Generator Sweep Signal

(3) View the swept frequency signal
Connect the output of the PM5193 to CH1 of the scope. Be careful not to change the triggering from the SWEEP on CH2. You should see a waveform similar to that shown in Figure 4.2.

(4) View the frequency response of a circuit
Connect the function generator output to the circuit input and connect the circuit output to CH1 of the scope. For a low pass filter circuit, the result would resemble the response shown in Figure 4.4. The cutoff frequency (f_c) can be found by estimating where the amplitude reaches 0.707, and by determining the frequency value that corresponds to that point in the sweep. For example, if cutoff point is at 1/5th of the distance from the start of the sweep to the end, f_c would be 800 Hz (4000

Hz / 5). You may use the oscilloscope cursors to estimate the cutoff frequency.

An alternative method to estimate the cutoff frequency is to manually select individual input frequencies on the function generator without using the sweep function. Start with a low frequency and increment the frequency (with the same amplitude) until the output amplitude on the scope reduces to 0.707 of the input value.

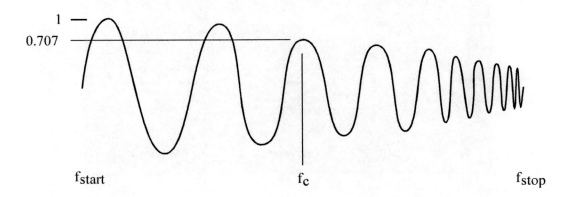

Figure 4.4 Sweep Frequency Response

4.4   Laboratory Procedure / Summary Sheet

Group: _____   Names: _____

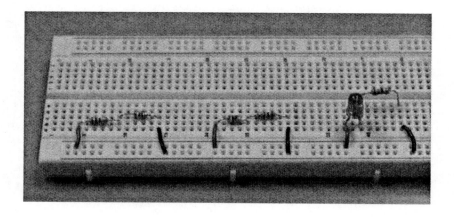

Figure 4.5 Circuits for Steps 1 and 3

(1) Build each filter circuit shown below. For each, use the procedures outlined in the previous section to generate the frequency sweep response and to estimate the cutoff frequency. Also, make a rough sketch of the response.

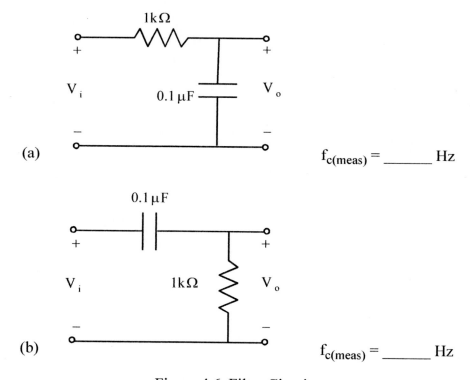

(a) $f_{c(meas)} =$ _____ Hz

(b) $f_{c(meas)} =$ _____ Hz

Figure 4.6 Filter Circuits

58

(2) Examine the silicon diode and LED. Decide which lead is the anode and which is the cathode. As shown in the Figure 4.7, the anode lead on an LED is longer. Using the 34401 DMM, select the diode test function (▷⊢) and determine whether or not your assumptions about the anodes were correct. You will notice that the diode test does not work properly for the LED. This is because the LED voltage drop is larger than the expected range for a silicon diode (0.3 V to 0.8 V). However, you should see the LED light up when properly biased by the DMM. Write down the measured voltage drop across the silicon diode.

$$V_{diode} = \underline{\hspace{2cm}} V$$

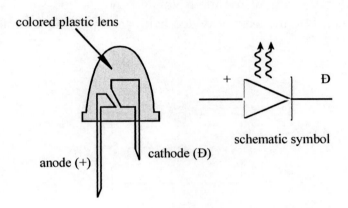

Figure 4.7 LED

(3) Construct the circuit in Figure 4.8 for both the diode and the LED and record the indicated voltages. Make a sketch for each output voltage superimposed on the input for the signals labeled with an asterisk below.

V_i	V_D (diode)	V_o(diode)	V_D (LED)	V_o(LED)
+5 V				
-5 V				
2 sin (6πt)	*	*	*	*

*: Sketch one cycle of the input voltage and the measured voltages (V_D and V_o) versus time. Use the axes provided below and provide appropriate scales and label each curve.

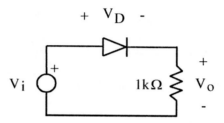

Figure 4.8 Diode/LED Circuit

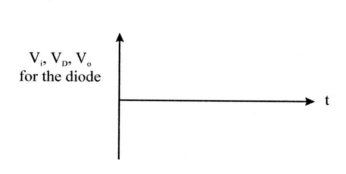

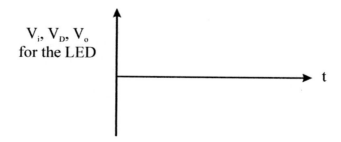

60

LAB 4 QUESTIONS

Group: _____   Names: _____

(1)   For the circuit in Step 1a:

what type of filter is this? _____

Also, calculate the theoretical cutoff frequency and the percent error in your measured value:

$\omega_{c(meas)} =$ _____ $\omega_{c(theor)} =$ _____   % error = _____

Remember that $\omega = 2\pi f$.

(2)   For the circuit in Step 1b, derive expressions for the magnitude ratio of the frequency response and for the cutoff frequency (ω_c).

$\left| \dfrac{V_o}{V_i} \right| =$ $\omega_c =$

what type of filter is this? _____

Also, calculate the theoretical cutoff frequency and the percent error in your measured value:

$\omega_{c(meas)} = $ _____ $\omega_{c(theor)} = $ _____   % error = _____

(3)   In step 3, if the diode where removed and replaced in the opposite direction in the circuit, what effect would this have on the outputs for the sin wave inputs?

Laboratory 5

Transistor and Photoelectric Circuits

Required Components:
- 2 330Ω resistors   or, or, brown
- 2 1 kΩ resistors
- 1 10kΩ resistor   brown black orange
- 1 20kΩ trim potentiometer
- 1 2N3904 small signal transistor
- 1 TIP31C power transistor
- 1 power diode
- 1 Radio Shack 1.5-3V DC motor (RS part number: 273-223)
- 1 LED
- 1 photodiode/phototransistor pair (Digikey part number: H21A1QT-ND)

5.1   Objectives

In this laboratory, you will study bipolar junction transistors (BJTs) and common photoelectric components. You will learn how to use light-emitting diodes (LEDs) as indicators, switch an inductive load with a power BJT, and use LED and phototransistor pairs as photo-interrupters. You will also learn how to bias a transistor and how to provide flyback protection with a diode.

5.2   Introduction

The following two pages provide information from the 2N3904 transistor data sheet. Data sheets provide pin-out information, where each pin is labeled with a function name and, if appropriate, a number. A data sheet also provides detailed electrical specifications that can help you properly design a circuit using the component.

Figure 5.1 illustrates the nomenclature used to describe the behavior of an npn bipolar transistor. It is a three terminal device consisting of the base, collector, and emitter. The transistor acts like a current valve by using the voltage bias across the base and emitter (V_{BE}) to control the flow of current in the collector-emitter circuit (I_C). The circuit connected to the collector and emitter along with the bias voltage dictate how much current flows.

2N3904   **MMBT3904**   **PZT3904**

TO-92

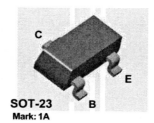

SOT-23
Mark: 1A

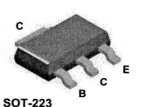

SOT-223

NPN General Purpose Amplifier

This device is designed as a general purpose amplifier and switch.
The useful dynamic range extends to 100 mA as a switch and to
100 MHz as an amplifier.

Absolute Maximum Ratings* T_A = 25°C unless otherwise noted

Symbol	Parameter	Value	Units
V_{CEO}	Collector-Emitter Voltage	40	V
V_{CBO}	Collector-Base Voltage	60	V
V_{EBO}	Emitter-Base Voltage	6.0	V
I_C	Collector Current - Continuous	200	mA
T_J, T_{stg}	Operating and Storage Junction Temperature Range	-55 to +150	°C

*These ratings are limiting values above which the serviceability of any semiconductor device may be impaired.

NOTES:
1) These ratings are based on a maximum junction temperature of 150 degrees C.
2) These are steady state limits. The factory should be consulted on applications involving pulsed or low duty cycle operations.

Thermal Characteristics T_A = 25°C unless otherwise noted

Symbol	Characteristic	Max			Units
		2N3904	*MMBT3904	**PZT3904	
P_D	Total Device Dissipation Derate above 25°C	625 5.0	350 2.8	1,000 8.0	mW mW/°C
$R_{\theta JC}$	Thermal Resistance, Junction to Case	83.3			°C/W
$R_{\theta JA}$	Thermal Resistance, Junction to Ambient	200	357	125	°C/W

*Device mounted on FR-4 PCB 1.6" X 1.6" X 0.06."

**Device mounted on FR-4 PCB 36 mm X 18 mm X 1.5 mm; mounting pad for the collector lead min. 6 cm².

NPN General Purpose Amplifier
(continued)

Electrical Characteristics $T_A = 25°C$ unless otherwise noted

Symbol	Parameter	Test Conditions	Min	Max	Units

OFF CHARACTERISTICS

Symbol	Parameter	Test Conditions	Min	Max	Units
$V_{(BR)CEO}$	Collector-Emitter Breakdown Voltage	$I_C = 1.0$ mA, $I_B = 0$	40		V
$V_{(BR)CBO}$	Collector-Base Breakdown Voltage	$I_C = 10$ μA, $I_E = 0$	60		V
$V_{(BR)EBO}$	Emitter-Base Breakdown Voltage	$I_E = 10$ μA, $I_C = 0$	6.0		V
I_{BL}	Base Cutoff Current	$V_{CE} = 30$ V, $V_{EB} = 3V$		50	nA
I_{CEX}	Collector Cutoff Current	$V_{CE} = 30$ V, $V_{EB} = 3V$		50	nA

ON CHARACTERISTICS*

Symbol	Parameter	Test Conditions	Min	Max	Units
h_{FE}	DC Current Gain	$I_C = 0.1$ mA, $V_{CE} = 1.0$ V	40		
		$I_C = 1.0$ mA, $V_{CE} = 1.0$ V	70		
		$I_C = 10$ mA, $V_{CE} = 1.0$ V	100	300	
		$I_C = 50$ mA, $V_{CE} = 1.0$ V	60		
		$I_C = 100$ mA, $V_{CE} = 1.0$ V	30		
$V_{CE(sat)}$	Collector-Emitter Saturation Voltage	$I_C = 10$ mA, $I_B = 1.0$ mA		0.2	V
		$I_C = 50$ mA, $I_B = 5.0$ mA		0.3	V
$V_{BE(sat)}$	Base-Emitter Saturation Voltage	$I_C = 10$ mA, $I_B = 1.0$ mA	0.65	0.85	V
		$I_C = 50$ mA, $I_B = 5.0$ mA		0.95	V

SMALL SIGNAL CHARACTERISTICS

Symbol	Parameter	Test Conditions	Min	Max	Units
f_T	Current Gain - Bandwidth Product	$I_C = 10$ mA, $V_{CE} = 20$ V, $f = 100$ MHz	300		MHz
C_{obo}	Output Capacitance	$V_{CB} = 5.0$ V, $I_E = 0$, $f = 1.0$ MHz		4.0	pF
C_{ibo}	Input Capacitance	$V_{EB} = 0.5$ V, $I_C = 0$, $f = 1.0$ MHz		8.0	pF
NF	Noise Figure	$I_C = 100$ μA, $V_{CE} = 5.0$ V, $R_S = 1.0$kΩ, $f=10$ Hz to 15.7kHz		5.0	dB

SWITCHING CHARACTERISTICS

Symbol	Parameter	Test Conditions	Min	Max	Units
t_d	Delay Time	$V_{CC} = 3.0$ V, $V_{BE} = 0.5$ V,		35	ns
t_r	Rise Time	$I_C = 10$ mA, $I_{B1} = 1.0$ mA		35	ns
t_s	Storage Time	$V_{CC} = 3.0$ V, $I_C = 10$mA		200	ns
t_f	Fall Time	$I_{B1} = I_{B2} = 1.0$ mA		50	ns

*Pulse Test: Pulse Width ≤ 300 μs, Duty Cycle ≤ 2.0%

Spice Model

NPN (Is=6.734f Xti=3 Eg=1.11 Vaf=74.03 Bf=416.4 Ne=1.259 Ise=6.734 Ikf=66.78m Xtb=1.5 Br=.7371 Nc=2
Isc=0 Ikr=0 Rc=1 Cjc=3.638p Mjc=.3085 Vjc=.75 Fc=.5 Cje=4.493p Mje=.2593 Vje=.75 Tr=239.5n Tf=301.2p
Itf=.4 Vtf=4 Xtf=2 Rb=10)

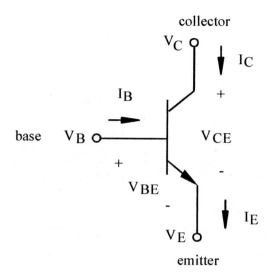

Figure 5.1 npn Bipolar Transistor Symbol and Nomenclature

Here are some general relationships between the variables shown in Figure 5.1:

$$V_{BE} = V_B - V_E \tag{5.1}$$

$$V_{CE} = V_C - V_E \tag{5.2}$$

$$I_E = I_B + I_C \tag{5.3}$$

Also, generally,

$$V_C > V_E \tag{5.4}$$

When the transistor is in saturation (i.e., fully ON),

$$V_{BE} \approx 0.6V \text{ to } 0.7V, \quad V_{CE} \approx 0.2V, \text{ and } I_C \gg I_B \tag{5.5}$$

and when the transistor is in its cutoff state,

$$V_{BE} < 0.6V \quad \text{and} \quad I_B = I_C = I_E = 0 \tag{5.6}$$

In the cutoff state, the transistor does not conduct current.

5.3   Laboratory Procedure / Summary Sheet

Group: ____   Names: _____

(1)   Build the simple LED indicator circuit shown below (without the 2nd resistor).
Gradually increase V_{in} from 0 V to 5 V and measure V_D when you consider the LED
to be on.

V_D (on) = __2,5__ *lingh*

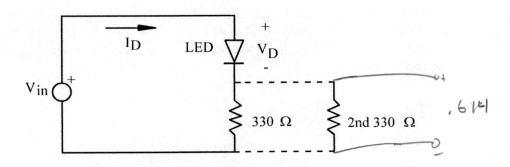

.6141

Figure 5.2 LED Circuit

Add the second resistor and explain what happens and why.

The LED emitts more light. The resistance of the circuit
has been reduced / allowing more current to flow through
the LED.

Calculate I_D.

Assuming zero resistance in LED   NO

$$I = \frac{V}{R} = \frac{2,5v}{165\Omega} = .015 \text{ Amps}$$

$I_D = \frac{V_{in} - V_D}{R_{eq}}$

(2) Sketch a voltage divider circuit that uses the potentiometer and a 10 V power source. Build the circuit and verify that it works properly by adjusting it to produce an output of 2.5V.

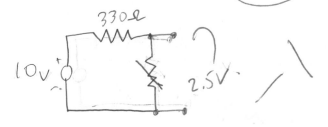

(3) Build a simple transistor switch (see figure below) using a 2N3904 small signal transistor and a base resistor (R_B) of 1 kΩ. Use the function generator for V_{in} so it can be later adjusted in small increments. Use the DC power supply for the 10V source.

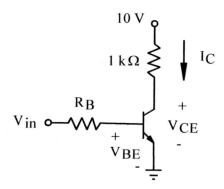

Figure 5.3 Transistor Switch

Use the 2N3904 datasheet provided in Section 5.2 to help you draw and label the pins on the figure below and to record the following values:

$I_{C(max)}$ = _____ $V_{C(max)}$ = _____

$V_{BE}(sat)$ = _____

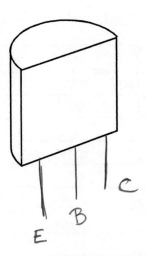

10v

Figure 5.4 2N3904 Pin-out

Vary V_{in} as indicated in the table below and record the associated values for V_{BE} and V_{CE}. Use $R_B = 1 \text{ k}\Omega$ for the base resistor.

V_{in}	V_{BE}	V_{CE}
0.0	.015	10.013
0.4	.467	10.013
0.5	.524	10.007
0.6	.662	9.669
0.7	.701	8.275
0.8	.785	1.219
0.9	.758	.184
1.0	.760	.159

Describe your conclusions about when saturation occurs for the transistor.

The transistor saturates at about Vin = .8v

69

Change the base resistor (R_B) to 10 kΩ and repeat the measurements.

V_{in}	V_{BE}	V_{CE}
0.0	.015	9.988
0.4	.417	9.991
0.5	.505	9.992
0.7	.690	8.875
0.9	.716	6.759
1.1	.735	3.421
1.3	.748	.724
1.5	.754	.240

What is the effect of a larger base resistor? Why?

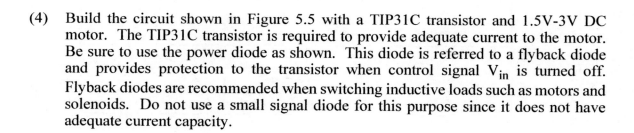

(4) Build the circuit shown in Figure 5.5 with a TIP31C transistor and 1.5V-3V DC motor. The TIP31C transistor is required to provide adequate current to the motor. Be sure to use the power diode as shown. This diode is referred to a flyback diode and provides protection to the transistor when control signal V_{in} is turned off. Flyback diodes are recommended when switching inductive loads such as motors and solenoids. Do not use a small signal diode for this purpose since it does not have adequate current capacity.

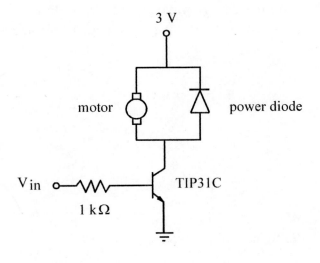

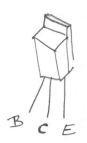

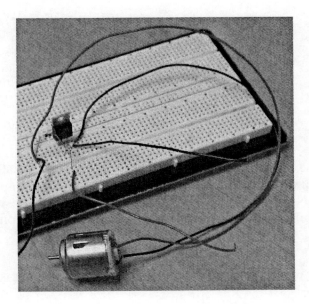

Figure 5.5 Motor and Flyback Diode

Gradually increase V_{in} from 0 V to 5 V and describe what happens.

The motor starts to turn at about 3v
and rpm increase with voltage.

Apply a 5 V amplitude square wave input to V_{in}. Start with a low frequency (e.g., 1 Hz) and then try some higher frequencies (e.g., 10 Hz and 100 Hz). Describe what happens.

motor pulses at low frequency and smooths out at higher frequency

Explain how the flyback diode works.

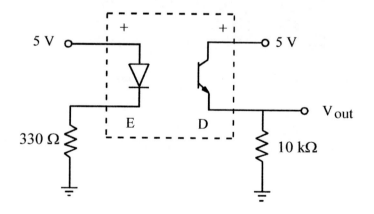

(5) Examine the photo-interrupter and look at its specifications. Explain why the indicated values for the resistors shown in Figure 5.6 are appropriate, and then build the circuit. Note that a single 5V source can be used to provide both voltage signals, and the ground for the input and output circuits must be connected to be common.

Figure 5.6 Photo-interrupter

Measure the output voltage (V_{out}) with and without the beam interrupted (e.g., by a sheet of paper). What conditions (interrupted or not) correspond to the high and low states of the output? Explain why each condition results in the respective state.

Why are the resistors required?

Laboratory 6

Operational Amplifier Circuits

Required Components:
- 1 741 op amp
- 2 1kΩ resistor
- 4 10kΩ resistors
- 1 100kΩ resistor
- 1 0.01μF capacitor

6.1   Objectives

Operational amplifiers are one of the most commonly used circuit elements in analog signal processing. Because of their wide range of applications you should become familiar with the basic terminal characteristics of operational amplifiers and the simple, yet powerful circuits that can be built with a few additional passive elements.

In this laboratory exercise you will examine a few of the electrical parameters that are important in the design and use of circuits containing operational amplifiers. These parameters will illustrate how the real operational amplifier differs form the ideal op amp which we have discussed in class. These parameters are:

    (1)   the input impedance

    (2)   the output voltage swing

    (3)   the slew rate

    (4)   the gain-bandwidth product

Also during this laboratory exercise you will construct and evaluate the performance of the following operational amplifier circuits:

    (1)   A non-inverting amplifier

    (2)   An inverting amplifier

    (3)   a voltage follower

    (4)   an integrator

    (5)   a differential amplifier

Figure 6.1 represents the basic model for an amplifier. The model assumes a differential input, an input impedance between the two input connections, and a dependent voltage source with gain A and series output impedance. This model can be used to develop the terminal characteristics of an operational amplifier.

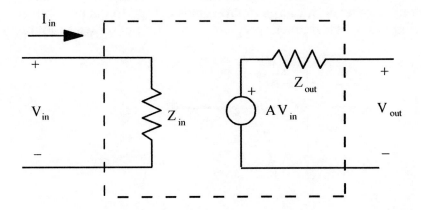

Figure 6.1 Amplifier Model

First, let the input impedance approach infinity and note what happens to the input current I_{in}:

$$\text{as } Z_{in} \rightarrow \infty, I_{in} \rightarrow 0 \qquad (6.1)$$

Thus, an ideal operational amplifier, assumed to have infinite input impedance, draws no current.

Now, let the gain A of the dependent source approach infinity as the output voltage (V_{out}) remains constant and note what happens to the input voltage V_{in}.

$$\text{as } A \rightarrow \infty, V_{in} \rightarrow 0 \qquad (6.2)$$

When an ideal operational amplifier, assumed to have infinite gain, is used in a circuit with negative feedback, the voltage difference between the input terminals is zero.

These ideal terminal characteristics greatly simplify the analysis of electrical networks containing operational amplifiers. They are only approximately valid, however.

Real operational amplifiers have terminal characteristics similar to those of the ideal op amp. They have a very high input impedance, so that very little current is drawn. At the same time, there is very little voltage drop across the input terminals. However, the input impedance of a real op amp is not infinite and its magnitude is an important terminal characteristic of the op amp. The gain of a real op amp is very large (100,000 or above), but not infinite.

Another important terminal characteristic of any real op amp is related to the maximum output voltage that can be obtained from the amplifier. Consider a non-inverting op amp circuit with a gain of 100 set by the external resistors. For a one volt input you would expect a 100 V output. **In reality, the maximum voltage output will be about 1.4 V less than the supply voltage to the op amp (V_{cc}) for infinite load impedance.**

Two other important characteristics of a real op amp are associated with its response to a square wave input. Ideally, when you apply a square wave input to an op amp you would expect a square wave output. However, for large input signals at high frequencies, deviations occur. The response of an op amp to a high frequency square wave input is shown in Figure 6.2.

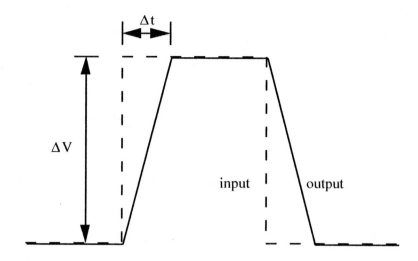

Figure 6.2 Effect of Slew Rate on a Square Wave

In order to quantify the response shown above, two operational amplifier parameters are defined:

Slew Rate: The maximum time rate of change of the output voltage

$$SR = \left(\frac{\Delta V}{\Delta t}\right)_{max} \quad (6.3)$$

Rise Time: The time required for the output voltage to go from 10% to 90% of its final value. This parameter is specified by manufacturers for specific load input parameters.

Another important characteristic of a real op amp is its frequency response. An ideal op amp exhibits infinite bandwidth. In practice, real op amps have a finite bandwidth which is a function of the gain set by external components. This gain is called the closed loop gain.

To quantify this dependence of bandwidth on the gain another definition is used, the Gain-Bandwidth Product (GBP). The GBP of an op amp is the product of the open loop gain and the bandwidth at that gain. The GBP is constant over a wide range of frequencies due to the linear relation shown in the log-log plot in Figure 6.3. The curve in the figure represents the maximum open loop gain of the op amp (where no feedback is included) for different input frequencies. The bandwidth of an op amp circuit with feedback will be limited by this open loop gain curve. Once the gain is selected by the choice of feedback components, the bandwidth of the resulting circuit extends from DC to the intersection of the gain with the open loop gain curve. The frequency at the point of intersection is called the fall-off frequency because the gain decreases logarithmically beyond this frequency. For example, if a circuit has a closed loop gain of 10, the fall-off frequency would be approximately 10,000 (10^5).

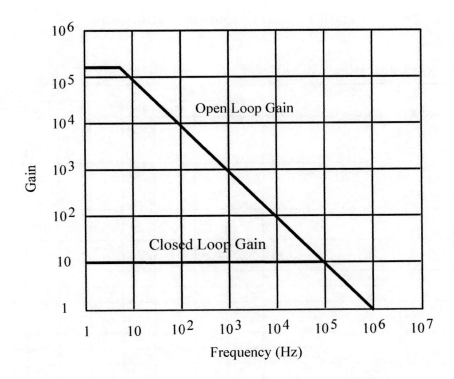

<u>Figure 6.3</u> Typical Open Loop Gain vs. Bandwidth for 741 Op Amp

Figure 6.4 shows the pin-out diagram and schematic symbol from the LM741 Op Amp datasheet. Table 6.1 shows some of the important electrical specifications available in the datasheet. The complete datasheet can be found in manufacturer linear circuits handbooks.

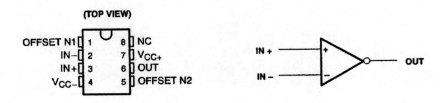

<u>Figure 6.4</u> LM741 Pin-out Diagram and SchematicSymbol

Table 6.1 LM741 Electrical Specifications

Electrical Characteristics (Note 5)

Parameter	Conditions	LM741A			LM741			LM741C			Units
		Min	Typ	Max	Min	Typ	Max	Min	Typ	Max	
Input Offset Voltage	$T_A = 25°C$										
	$R_S \leq 10\ k\Omega$					1.0	5.0		2.0	6.0	mV
	$R_S \leq 50\Omega$		0.8	3.0							mV
	$T_{AMIN} \leq T_A \leq T_{AMAX}$										
	$R_S \leq 50\Omega$			4.0							mV
	$R_S \leq 10\ k\Omega$						6.0			7.5	mV
Average Input Offset Voltage Drift				15							µV/°C
Input Offset Voltage Adjustment Range	$T_A = 25°C$, $V_S = \pm20V$	±10				±15			±15		mV
Input Offset Current	$T_A = 25°C$		3.0	30		20	200		20	200	nA
	$T_{AMIN} \leq T_A \leq T_{AMAX}$			70		85	500			300	nA
Average Input Offset Current Drift				0.5							nA/°C
Input Bias Current	$T_A = 25°C$		30	80		80	500		80	500	nA
	$T_{AMIN} \leq T_A \leq T_{AMAX}$			0.210			1.5			0.8	µA
Input Resistance	$T_A = 25°C$, $V_S = \pm20V$	1.0	6.0		0.3	2.0		0.3	2.0		MΩ
	$T_{AMIN} \leq T_A \leq T_{AMAX}$, $V_S = \pm20V$	0.5									MΩ
Input Voltage Range	$T_A = 25°C$							±12	±13		V
	$T_{AMIN} \leq T_A \leq T_{AMAX}$				±12	±13					V

6.2   Laboratory Procedure / Summary Sheet

Group: ____   Names: _____

NOTE - An op amp always requires connection to an external power supply. Usually, two DC power supplies are required: +15V and -15V. The two different voltage levels can be conveniently provided by a triple output power supply. When you build an op amp circuit, always check that a power source is connected to the op amp.

(1)   We will examine the usefulness of the high input impedance of the op amp by constructing the simple circuit known as the voltage follower. Begin by building the circuit shown in Figure 6.5a consisting of a voltage divider (R_1, R_2) and a load resistance (R_3) where $R_1=R_2=R_3=10k\Omega$. Calculate the expected values for V_1 and V_{out} and measure them:

voltage	calculated	measured
V_1		
V_{out}		

Then insert the op amp follower between the voltage divider and the load resistor as shown in Figure 6.5b. Be sure the op amp has the proper power supply connections as well as the signal connections shown in the figure. Again calculate the expected values for V_1 and V_{out} and measure them:

voltage	calculated	measured
V_1		
V_{out}		

Explain the differences among the voltages measured in the two circuits.

You should be able to see now that the follower isolates the left part of the circuit from the right part. The follower effectively changes a high impedance output to a low impedance output. The result is that the output of the voltage divider is not changed by different load resistors. Explain why.

79

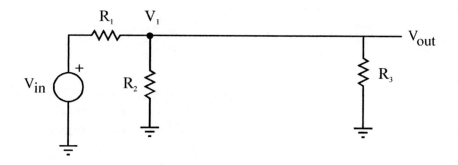

(a) without op amp follower

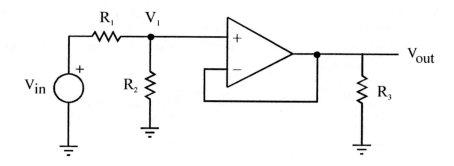

(b) with op amp follower

Figure 6.5 Voltage Divider Driving Load Resistor

(2) Construct an inverting amplifier with a gain of -10 and use it to determine the maximum swing voltage in the following way. First, apply a 1 V_{pp} 1kHz sinusoidal signal. Then, increase the amplitude of the input slowly and note where the sinusoidal output is first distorted as you increase the input voltage. Be sure to use resistors in the kΩ range. Consider the input and output currents to explain why large resistance values are necessary.

(3) Construct the modified integrator shown below. Normally, the shunt resistor (R_s) is selected such that $R_s \geq 10\ R_1$. Also, the product R_1C is chosen to be approximately equal to the period of the applied input voltage signal. Apply a 10 KHz, 1 V_{p-p} square wave. Use the following component values: $C = 0.01\ \mu F$, $R_s = 100\ k\Omega$, and $R_1 = R_2 = 10\ k\Omega$. Justify these selections.

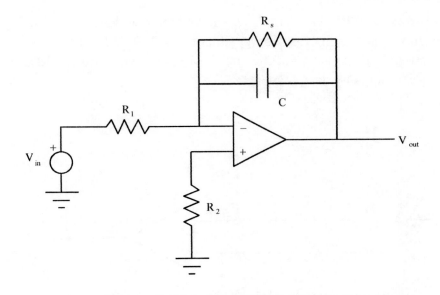

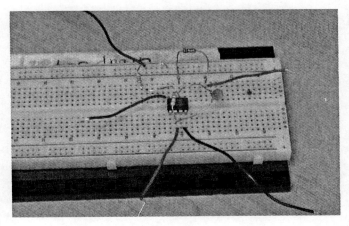

Figure 6.6 Integrator

(4)   For the circuit above, determine experimentally the frequency range over which the circuit functions as an integrator. To do this systematically, adjust the input signal to be a 1 V_{pp} square-wave with no DC offset. As you vary the frequency over a wide range you will notice that the output will deviate from the expected triangular wave (integrated square wave). Determine and report the frequency below which the circuit does not operate as an integrator.

(5)   Construct the difference amplifier shown below with a gain of 1 using $R_1=R_F=10k\Omega$. Set the output of the function generator to a 1 V_{pp}, 1 kHz sine wave and attach the output to both inputs V_s of the circuit. Explain what you would expect at the output V_{out} and note any discrepancies.

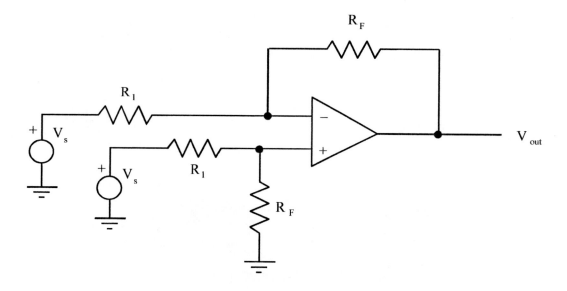

Figure 6.7 Difference Amplifier

LAB 6 QUESTIONS

Group: _____ Names: _____

(1) Find the specifications for the 741 op amp in the Linear Data Book. Record the values for each of the characteristic parameters listed below. Also, discuss the significance of each parameter.

- input impedance

- output impedance

- open loop gain

- swing voltage

- short circuit output current

(2) Explain why the voltage follower isolates the input from the output.

(3) What is the fall-off frequency (approximate bandwidth) of a 741 op amp circuit designed with a closed loop gain of 100?

(4) The output of the difference amp was not exactly zero when the inputs are of equal magnitudes. Suggest possible causes for this discrepancy.

Laboratory 7

Digital Circuits - Logic and Latching

Required Components:
- 1 330Ω resistor
- 4 1kΩ resistor
- 2 0.1μF capacitor
- 1 2N3904 small signal transistor
- 1 LED
- 1 7408 AND gate IC
- 1 7474 positive edge triggered flip-flop IC
- 1 7475 data latch IC
- 3 SPDT switches or NO buttons

7.1   Objectives

In this laboratory exercise you will use TTL (transistor-to-transistor logic) integrated circuits (ICs) to perform combinational and sequential logic functions. Specifically, you will learn how to use logic gates and flip-flops. You will use these components to build a simple circuit to control the display of an LED based on the past and current state of various switches or buttons. You will also learn how to read manufacturer TTL data books that summarize the functionality and specifications for a whole family of digital devices.

7.2   Introduction

The ICs you will be handling in this laboratory exercise require digital inputs and produce digital outputs. A binary digital signal is a sequence of discrete states, in contrast to an analog signal that varies continuously. Figure 7.1 shows the difference between digital and analog signals. The sampled digital data is a discrete representation of the analog signal. The data is represented by a series of bits.

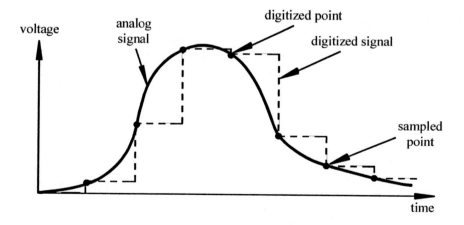

Figure 7.1 Analog and Digital Signals

A binary (digital) signal may exist in only one of two states defined as a voltage high and low. Many types of devices are available for processing the information contained within a digital signal (i.e., a sequence of 0s and 1s). The ICs you will see in this laboratory exercise are TTL (Transistor-Transistor Logic) circuits. TTL devices process digital signals that have a high level defined from 3.5 to 5 V and a low level between 0 to 0.7 V. 0.7 V to 3.5 V is a dead zone. Usually, but not always, a voltage high is equivalent to a logic high. Each of the signals at the input and output terminals of a digital device can exist in only one of two possible states, a voltage low corresponding to a binary zero, or a voltage high corresponding to a binary one.

There are hundreds of TTL ICs (also called chips) available, each with its own functionality. There are many companies that manufacture ICs, but they all use a standard numbering method to identify the ICs. Each chip manufacturer publishes a set of data books that describe how each of the ICs work. In the Lab, we have TTL data books from National Semiconductor, Texas Instruments, and Motorola. These books all contain the same basic information: chip pin-outs, truth tables, operating ranges, and chip-specific details. At the front of each book is a functional index that lists all the chips described in the book according to their function. This is the first place you should look when trying to find a chip for a particular application. For example, let's say you need an AND gate (as you do for this exercise). The Motorola Data book lists a "Quad 2-Input AND Gate" with a device number of MC54/74F08. The National Semiconductor data book also lists a "Quad 2-Input AND Gate," but with a device number of DM74LS08. For most purposes, the only numbers that are important are the "74," which corresponds to the standard TTL series, and the "08," which identifies the chip function (in this case, a Quad 2-Input AND). A standard "Quad 2-Input AND Gate" can be referred to simply as a "7408" for any manufacturer. The information in the data book in organized in numerical order according to the chip unique number (in this case, 08). Knowing the chip number from the functional index, you can now find the chip information in the book.

7.3   Data Flip-flops and Latches

There are many digital circuit applications where you may need to store data for later use. One way to do this is through the use of flip-flops. The bistable data latch (see Figure 7.2) is a flip-flop that is useful in many applications. The data latch has a data input (D), a clock input (CK), and output Q. Most flip-flops include complementary outputs where $\overline{Q}$ is the inverse of Q. With a data latch, the data input gets passed to or blocked from the output depending upon the clock signal. When CK is high, Q=D (i.e., the output tracks the input). When CK is low, the D input is ignored and the last value of Q (the value of D when CK last went low) is stored (latched). This memory state of the flip-flop (when CK is low) is indicated in the truth table with Q_0 (the last value latched). The entire functionality is summarized in the truth table shown in the figure. An X in a truth table indicates that a signal value may have either value (H or L). For example, for the data latch, when CK is low, the input D has no effect on the output Q.

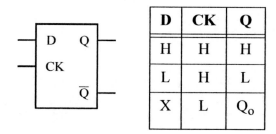

D	CK	Q
H	H	H
L	H	L
X	L	Q_o

Figure 7.2 Data Latch

The data latch is sometimes referred to as a level-triggered device since it is active (or triggered) based on the level (high or low) of the clock input, in this case high. A more common type of triggering for flip-flops is edge triggering where the output can change state only during a transition of the clock signal. Devices that respond when the clock transitions from low-to-high (indicated by an up arrow in a truth table) are referred to as positive edge triggered devices. Devices that respond when the clock transitions from high-to-low (indicated by a down arrow in a truth table) are referred to as negative edge triggered devices. Figures 7.3 and 7.4 summarize the functionality of positive and negative edge-triggered D-type flips-flops. Positive edge triggering is indicated by a triangle at the clock input. Negative edge triggering is indicated by an inversion circle and triangle at the clock input.

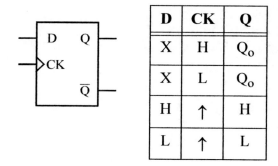

D	CK	Q
X	H	Q_o
X	L	Q_o
H	↑	H
L	↑	L

Figure 7.3 Positive Edge-triggered D flip-flop

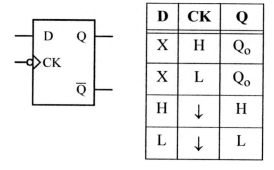

D	CK	Q
X	H	Q_o
X	L	Q_o
H	↓	H
L	↓	L

Figure 7.4 Negative Edge-triggered D flip-flop

Figures 7.5 through 7.7 show pin-out and schematic diagrams from the datasheets for various TTL devices used in this exercise.

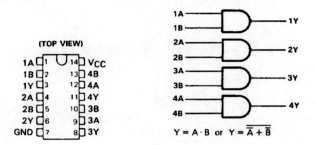

Figure 7.5 Pin-out and schematic symbol diagrams for the 7408

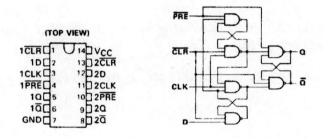

Figure 7.6 Pin-out and schematic symbol diagrams for the 7474

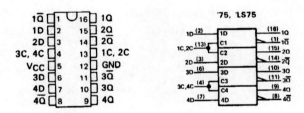

Figure 7.7 Pin-out and schematic symbol diagrams for the 7475

7.4   Hints on troubleshooting prototype circuits that involve integrated circuits

Please follow the protocol listed below when constructing and testing prototype circuits that use integrated circuits (ICs). Generally, if you carefully follow this protocol you will avoid a lot of frustration

(1)   Start with a clearly drawn schematic illustrating all components, inputs, outputs, and connections.

(2)   Draw a detailed wiring diagram, using the information from handbooks regarding device pin-outs. Label and number each pin used on each IC and fully specify each component. This will be your wiring guide.

(3)   Double check the functions you want to perform with each device.

(4)   Insert the ICs into your breadboard, and select appropriately colored wire (i.e. red for +5V, black for ground, other colors for signals).

(5)   Wire up all connections, overwriting the wiring diagram with a red pen or highlighter as you insert each wire.

(6)   Double check the +5V and ground connections to each IC.

(7)   Set the power supply to +5V and turn it off.

(8)   Connect the power supply to your breadboard and then turn it on.

(9)   Measure signals at inputs and outputs to verify proper functionality.

(10)   If your circuit is not functioning properly, go back through the above steps in reverse order checking everything carefully.

7.5   Laboratory Procedure / Summary Sheet

Group: _____   Names: _____

(1)   Use a TTL data book to look up the specifications for the 7408 Quad-AND, the 7474 positive edge-triggered flip-flop, and the 7475 bistable data latch. Pay particular attention to the pin-out diagrams. (Note - these are provided at the beginning of this Lab).

(2)   Using the datasheet pin-out diagrams, draw a complete, detailed wiring diagram for the circuit shown in Figure 7.9 using a 7474 positive edge-triggered flip-flop. Carefully label and number all pins used on each IC, including power and ground. Then construct the circuit. You will need to submit your detailed wiring diagram with your Lab summary and answered questions at the end of the Lab.

Note: it is good practice to include a 0.1 μF capacitor across the power and ground pins of each IC (not shown in Figure 7.9). This helps filter out transients that could occur on the power and ground lines during switching. The capacitors are especially important in more complicated circuits where a single power supply may be providing reference voltages and switched current to numerous components.

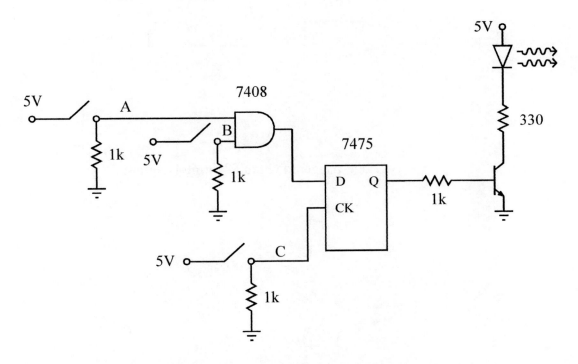

Figure 7.8 Circuit with Switches, Logic Gate, and Flip-flop

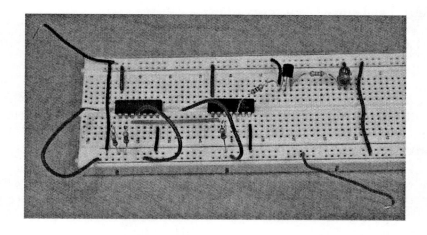

Figure 7.9 Photograph of the Partially Completed Circuit

(3)   Complete the following timing diagram and verify the results by testing your circuit.

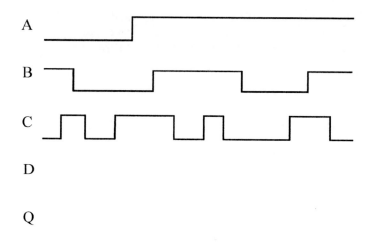

Figure 7.10 Edge-triggered Circuit Timing Diagram

(4) Replace the 7474 with the 7475 bistable data latch. Complete the following timing diagram and verify the results by testing your circuit.

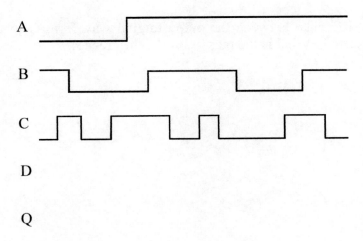

Figure 7.11 Latch Circuit Timing Diagram

LAB 7 QUESTIONS

Group: _____   Names: _____

(1)   Describe the difference between the output digital waveform Q in the two circuits you analyzed and tested. What is the reason for the difference?

(2)   The switches or buttons used exhibit bounce. Does this have an effect on the outputs Q of the circuits? If so, in what cases? Did you see the effects of bounce in the LED display? Why or why not?

(3)   What is the purpose for the resistors between the switch outputs and ground?

(4)   Attach the detailed wiring diagram that you used to construct your circuit.

Laboratory 8

Digital Circuits - Counter and LED Display

Required Components:
- 1 1kΩ resistor
- 1 10MΩ resistor
- 3 0.1μF capacitor
- 1 555 timer
- 1 7490 decade counter
- 1 7447 BDC to LED decoder
- 1 MAN 6910 or LTD-482EC seven-segment LED digital display
- 1 330Ω DIP resistor array

8.1   Objectives

In this laboratory exercise you will build a digital counter with a 1-digit decimal LED display. In doing so, you will learn to assemble and interconnect various integrated circuits to achieve sophisticated functionality.

8.2   Introduction

A common requirement in digital circuits applications is to count and display the number of pulses contained in a continuous TTL compatible pulse train (e.g., the output of a proximity sensor detecting parts on a moving conveyor belt or a photosensor detecting a reflection from a piece of tape on a rotating shaft). We want to count the number of pulses and output this number in binary coded form. This can be done using a 7490 decade counter. Refer to the 7490 pin-out and function information in Figure 8.1.

Connection Diagram

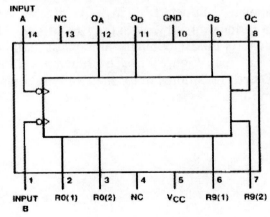

Reset/Count Truth Table

Reset Inputs				Output			
R0(1)	R0(2)	R9(1)	R9(2)	Q_D	Q_C	Q_B	Q_A
H	H	L	X	L	L	L	L
H	H	X	L	L	L	L	L
X	X	H	H	H	L	L	H
X	L	X	L	COUNT			
L	X	L	X	COUNT			
L	X	X	L	COUNT			
X	L	L	X	COUNT			

Figure 8.1 7490 Datasheet Information

The output of the counter is in binary coded decimal (BCD) form and consists of four bits, one bit presented by each of the four output terminals. The maximum number of combinations possible at the output of this device is 2^4 or 16. Each of the possible combinations and its corresponding output in BCD is shown in Table 8.1. Note that here a logic high corresponds to a voltage high. The BCD counters can then be cascaded in order to count in powers of 10.

Table 8.1 7490 Decade Counter BCD Coding

Decimal Count	Binary Code Output			
	Q_D	Q_C	Q_B	Q_A
0	0	0	0	0
1	0	0	0	1
2	0	0	1	0
3	0	0	1	1
4	0	1	0	0
5	0	1	0	1
6	0	1	1	0
7	0	1	1	1
8	1	0	0	0
9	1	0	0	1

The 7490 decade counter has four reset inputs: R0(1), R0(2), R9(1), and R9(2) that control count and reset functions. The Reset/Count Truth Table summarizing the functions of these four pins is included in Figure 8.1. There are many ways to utilize these reset inputs. A simple method is to set R0(2) = H, R9(1) = L, and R9(2) = L, where H=5V and L=0V. When R0(1) is set to L, the counter will be in count mode (see row 5 or 6 of the Reset/Count Truth Table in Figure 8.1). When R0(1) is set to H, the counter will reset to 0 (LLLL) (see row 2 of the Reset/Count Truth Table).

It is also convenient to display the output count on a 7 segment LED in digit form. Another device will be necessary to decode the four bits into a form compatible with the LED array. This device, the 7447 BCD-to-seven-segment decoder, converts the BCD binary number at its inputs into a 7 segment code to properly drive the LED digit (see Figure 8.2).The function table to describing the input (BCD) to output (7-segment LED code) relationship for the 7447 is shown in Table 8.2. Refer to Figure 8.3 for the pin-out diagram for the device.

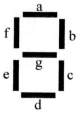

Figure 8.2 Seven-Segment LED Display (LCD)

94

Table 8.2 7447 BCD to 7-segment Decoder

Decimal Digit	Input				Output						
	Q_D	Q_C	Q_B	Q_A	a	b	c	d	e	f	g
0	0	0	0	0	0	0	0	0	0	0	1
1	0	0	0	1	1	0	0	1	1	1	1
2	0	0	1	0	0	0	1	0	0	1	0
3	0	0	1	1	0	0	0	0	1	1	0
4	0	1	0	0	1	0	0	1	1	0	0
5	0	1	0	1	0	1	0	0	1	0	0
6	0	1	1	0	1	1	0	0	0	0	0
7	0	1	1	1	0	0	0	1	1	1	1
8	1	0	0	0	0	0	0	0	0	0	0
9	1	0	0	1	0	0	0	1	1	0	0

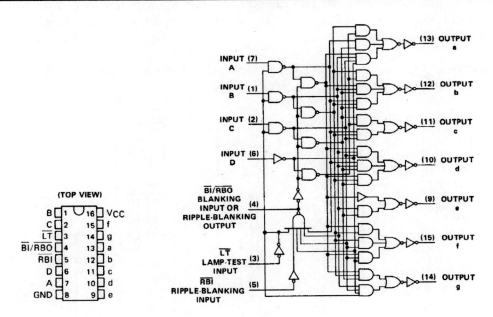

Figure 8.3 7490 Pin-out and Schematic Diagram

If the 7447 decoder driver is now properly connected to a 7 segment LED display, the count from the counter may be displayed in an easily recognizable form. It should be noted that the decoder driver does not actually drive the segment LEDs by supplying current to them, but by sinking current from them (see Figure 8.4). Therefore the LED is on when the 7447 output is low (0) allowing current to flow to ground.

95

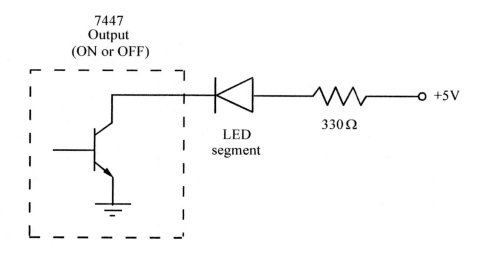

Figure 8.4 Output Circuit of 7447 and LED Driver

If the output transistor for a segment of the decoder driver is ON, then current flows through the LED and it lights. If on the other hand the transistor is OFF, no current flows and the LED is not lit. 330 ohm resistors are used to limit the current that is drawn by the decoder driver and to prevent BURNING OUT the LED.

As shown in Figure 8.5, the 7490 and single-digit LED displays can be cascaded to count and display any order of magnitude (10's, 100's, 1000's, etc.).

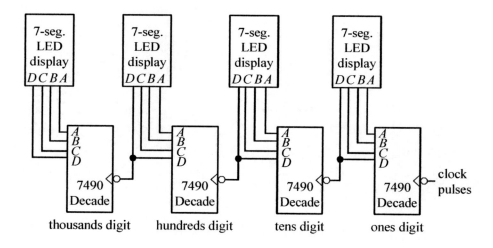

Figure 8.5 Cascading 7490s to display large count values

8.3 Procedure

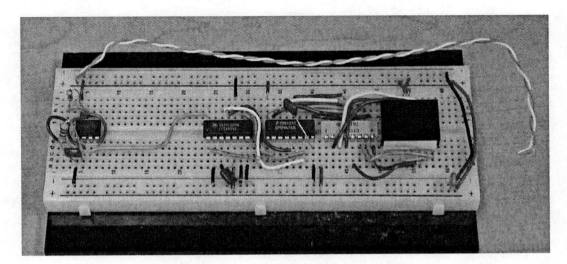

(1) Construct the timer circuit (see Figure 8.6) at one end of your protoboard carefully laying out the connections and wiring for easy debugging. Figure 8.7 shows information from the 555 datasheet. Using the components shown in Figure 8.6, the output of the circuit will be a pulse train with a frequency of approximately 0.7 Hz corresponding to a period of approximately 1.4 seconds.

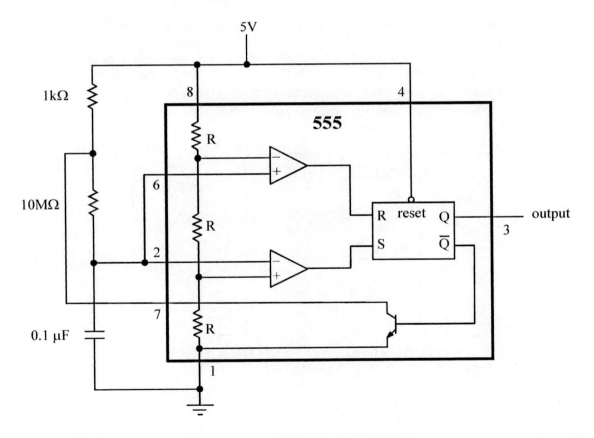

Figure 8.6 555 Timer Circuit

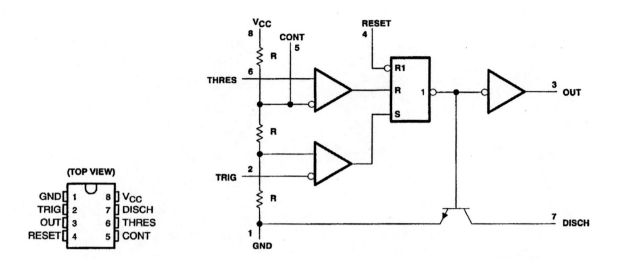

Figure 8.7 555 Pin-out and Schematic Diagram

Show the result to your Lab instructor and test the circuit to make sure it works properly. Leave this circuit on your protoboard as it will be used later.

(2) In the steps that follow, you will construct a one decade digital display as shown in Figure 8.8. Each group will be given an MAN6910 or LTD482EC LED digit display, a 7447, a 7490, and a 555. Carefully color code your wires, trim them to correct lengths, and insert them flat against the board. A "rat's nest" will not be acceptable. Please see the physical example provided by your TA. Figure 8.9 includes useful reference information from the MAN6910 datasheet.

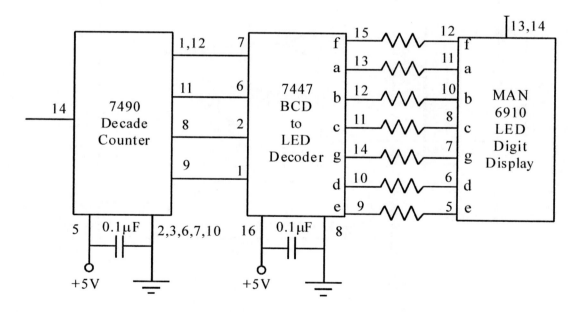

or

98

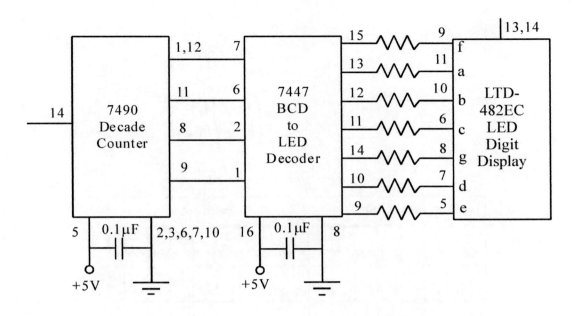

Figure 8.8 Decade Counter Circuit Schematic

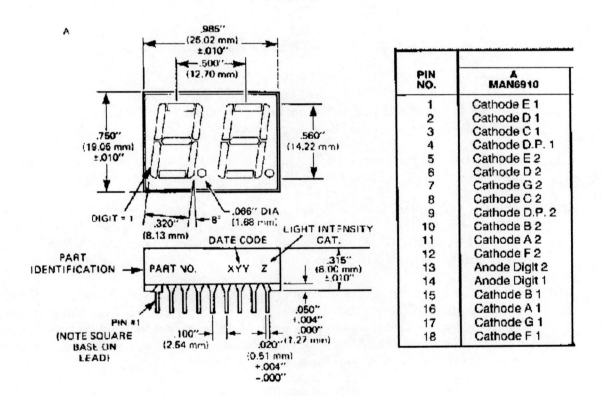

PIN NO.	A MAN6910
1	Cathode E 1
2	Cathode D 1
3	Cathode C 1
4	Cathode D.P. 1
5	Cathode E 2
6	Cathode D 2
7	Cathode G 2
8	Cathode C 2
9	Cathode D.P. 2
10	Cathode B 2
11	Cathode A 2
12	Cathode F 2
13	Anode Digit 2
14	Anode Digit 1
15	Cathode B 1
16	Cathode A 1
17	Cathode G 1
18	Cathode F 1

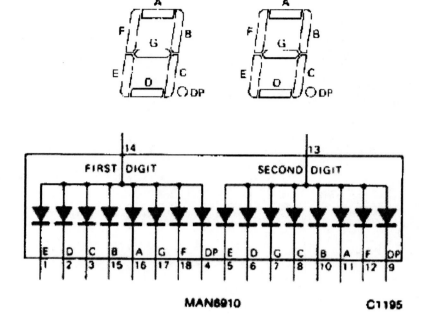

Figure 8.9 MAN6910 Datasheet Information

(3) Wire the top two rows of holes on the breadboard (with the long side oriented horizontally) to +5 V on the DC power supply.

(4) Wire the bottom two rows of holes on the breadboard to COMMON on your triple output power supply. These rows will constitute a bus for later wiring.

(5) Insert the MAN6910, 2-digit display, positioning it in the center of the protoboard with the corner labeled MAN6910 at the left-lower side. This is the reference for pin 1 on the MAN6910. Be sure to lay out all of the other components side-by-side to make sure they will fit on your protoboard.

(6) Insert the 7447 on one side of the display. Refer to the MAN6910 data sheet for pin-out information. Connect the MAN6910 to +5 V bus (pins 14, 13). Nothing more! Connect three 330 Ω resistors to pins 3, 15, 16 of the MAN6910 and ground the other ends of the resistors. Double-check your circuit, and then turn on the power supply. Is the displayed digit what you expected?

(7) Connect the MAN6910 ONE's (right) digit to its 7447 and then attach ground and +5V to the 7447.

(8) Attach the lamp test of the 7447 (pin 3, 0V and pin 4, 5V). Turn on the power supply and describe what happens. Note - all LED's should light.

(9) Remove the wire from pin 3 and apply +5V to pins 1, 2, 7 and GND pin 6 of the 7447. Describe the display.

(10) Write down in numerical order which pins of the 7447 should be high in order to display a 3.

(11) Finish wiring the 7447 and 7490 as shown in the schematic. Refer to the 7490 Reset/Count Truth Table at the beginning of the Lab and the description of the reset inputs in Section 8.2. In the schematic, all of the reset pins are grounded, putting the 7490 in count mode. If you wanted to be able to reset the counter (e.g., with at pushbutton input), you would need to set R0(2) = H, R9(1) = L, and R9(2) = L. In this case, if R0(1) is set to L, the counter would be in count mode and if R0(1) is switched to H, the counter would reset to 0.

(12) Attach the output of your 555 to the input of the 7490. Double-check your circuit! Turn on the power supply and describe your result.

(13) Demonstrate to the lab instructor that your display can increment properly from 0 to 9. At the same time, also demonstrate that you can reset the counter to 0!

LAB 8 QUESTIONS

Group: _____   Names: _____

(1)   Describe the results of the Procedure.

(2)   Which pins of the 7447 should be high to display a "b"?

(3)   How can you use a normally open (NO) button to reset the counter to 0? Draw a schematic.

Laboratory 9

Programming a PIC Microcontroller - Part I

Required Components:
- 1 PIC16F84 (4MHz) or PIC16F84A (20MHz) 18P-DIP microcontroller
- 1 4MHz microprocessor crystal (20 pF)
- 2 22pF capacitors
- 1 0.1 µF capacitor
- 1 LED
- 1 470Ω resistor
- 1 4.7kΩ resistor
- 1 SPST microswitch or NO button
- 1 1kΩ resistor

Required Special Equipment:
- Microchip PICSTART Plus development programmer
- Microchip's MPLAB integrated development environment Windows software
- MicroEngineering Labs' PicBasic Pro compiler

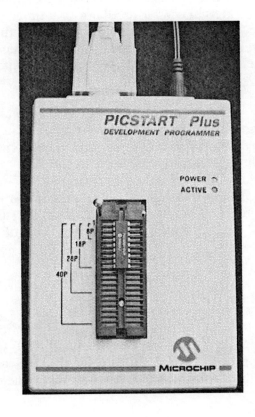

9.1   Objective

Microcontrollers are important parts in the design and control of mechatronic systems. This laboratory introduces the architecture of Microchip's PIC microcontroller, describes the PIC's capabilities, shows how to create programs using microEngineering Lab's PicBASIC Pro, and shows how to wire simple circuits using the PIC and your software. The exercise also shows how to use interrupts in response to sensor inputs to the PIC.

9.2   Introduction

A PIC microcontroller adds sophisticated digital control capabilities when connected to other circuits and devices. A single PIC microcontroller can communicate with other electronic devices, and digitally switch them on or off to control simple operations.

Microchip Technology, Inc. (*www.microchip.com*) produces a family of PIC processors capable of storing programs. The PIC16F84, which we will use in this Lab, contains electrically erasable programmable ROM (EEPROM), which is memory used to store programs. The program in EEPROM can be overwritten many times during the design cycle. The PIC has 64 bytes of data EEPROM and 1792 bytes of program EEPROM for storing compiled programs. It operates at 4 or 10 MHz depending upon an external crystal oscillator or a timer circuit. The PIC16F84A can function up to 20 MHz. The 18-pin PIC has 13 pins capable of operating as either inputs or outputs, designated by software, that can be changed during program execution. Five of the pins are grouped together and referenced as PORTA; another 8 pins are grouped together and referenced as PORTB.

Simple programs may be written in a form of BASIC called PicBasic Pro, which is available from microEngineering Labs, Inc. (*www.melabs.com*). The package includes a compiler that converts PicBasic to assembly language code, and then compiles the assembly code to hexadecimal machine code (hex) that is downloaded to the PIC. The hex executable code is downloaded via a serial port to a PIC using the Microchip Development Programmer hardware using a Windows interface. Once written, the program remains in PIC memory even when the power is removed. Example programs are included in the samples subdirectory of the PBP folder where PicBasic Pro is installed (e.g., C:\Program Files\PICBasicPro).

9.2.1   PIC Structure

The PIC16F84 is an 18-pin DIP IC with the pin-out shown in Figure 9.1. It has external power and ground pins (Vdd and Vss), 13 binary input/output (I/O) pins (RA[0-4] and RB[0-7]), and uses an external clock signal (OSC1 and OSC2). The master clear (MCLR) pin is active low, meaning the PIC is reset when the pin is grounded. MCLR must be held high during PIC operation.

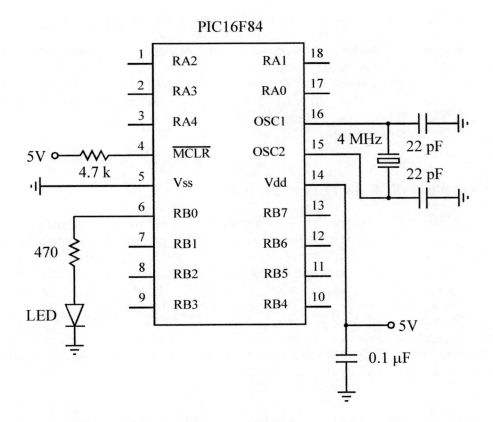

Figure 9.1 PIC and LED circuit schematic for blink.bas example

The power supply voltage (5 Vdc) and ground are connected to pins labeled Vdd and Vss (pins 14 and 5), respectively. Pin 4 (MCLR) is attached to 5Vdc with a 4.7K resistor to ensure continuous operation. If this pin were left unconnected (floating), the PIC could spontaneously reset itself. An accurate clock frequency can be obtained by connecting a 4MHz crystal (sometimes indicated as XT) across pins 15 and 16 which are also connected to ground through 22pF capacitors. A less expensive and less accurate alternative for setting a clock frequency is to attach an RC circuit to pin 16 while leaving pin 15 unattached (referred to as an RC clock).

The pins labelled RAx and RBx provide binary I/O. They are divided into two groups called PORTs. PORTA refers to pins RA0 through RA4 and PORTB refers to pins RB0 through RB7. PORTA and PORTB are compiler variable names that provide access to registers on the PIC. Each bit within the PORT can be referred to individually by its bit location (e.g., PORTA.3 refers to bit 3 in the PORTA register). For both ports, bit zero (PORTA.0 or PORTB.0) is the least significant bit (LSB). The specifics of how the PORT bits are defined and accessed follow:

PORTA: Designated in PicBasic Pro code as PORTA.0 through PORTA.4 (5 pins: 17, 18, and 1 through 3). For example, PORTA = %00010001 would set the PORTA.0 and PORTA.4 bits to 1, and set all other bits to 0. The % sign indicates binary number format. For PORTA, the three most significant bits are not required (i.e., %10001 would suffice).

PORTB: Designated in PicBasic Pro code as PORTB.0 through PORTB.7 (8 pins: 6 through 13). For example, PORTB = %01010001 would set PORTB.0,

PORTB.4, and PORTB.6 to 1. All other bits would be set to 0.

Each individual pin can be configured as an input or output independently (as described in the following Lab). When a pin is configured as an output, the output digital value (0 or 1) on the pin can be set with a simple assignment statement (e.g., PORTB.1 = 1). When a pin is configured as an input, the digital value on the pin (0 or 1) can be read by referencing the corresponding port bit directly (e.g., IF (PORTA.2 = 1) THEN ...).

9.3   An Example of PICBasic Pro Programming

PicBasic Pro is a compiler that uses a pseudocode approach to translate user friendly BASIC code into more cryptic assembly language code that is created in a separate *.asm file. The assembly code is then compiled into hexadecimal machine code (*.hex file), or hex code for short, that the PIC can interpret. The hex code file is then downloaded to the PIC and remains stored semi-permanently in EEPROM even when it is powered off. The code will remain in PIC memory until it is erased or overwritten using the Development Programmer.

For this laboratory you will program, compile, and test a very simple PicBasic example that controls the blinking of an LED. The code for this program, called blink.bas, is listed below. The hardware required is shown in Figure 9.1.

```
' blink.bas
' Example program to blink an LED connected to PORTB.0 about once a second

loop:   High PORTB.0              ' turn on LED connected to PORTB.0
        Pause 500                 ' delay for 0.5 seconds

        Low PORTB.0               ' turn off LED connected to PORTB.0
        Pause 500                 ' delay for 0.5 seconds

        Goto loop                 ' go back to label "loop" repeatedly
        End
```

The blink.bas program turns a light emitting diode (LED) on for half a second, and then turns it off for half a second, repeating the sequence for as long as power is applied to the circuit. The first two optional lines in the program are comment lines that identify the program and its function. Comment lines must begin with an apostrophe. On any line, information on the right side of an apostrophe is treated as a comment and is ignored by the compiler. The label "loop" allows the program to return control to this line at a later time using the Goto command. "High PORTB.0" causes pin 6 (RB0) to go high which turns on the LED. The Pause command delays execution of the next line of code by a given number of milliseconds (in this case 500 corresponds to 500 milliseconds or 0.5 second). "Low PORTB.0" causes pin 6 (RB0) to go low which turns the LED off. The following Pause causes a 500 millisecond delay before executing the next line. The "Goto loop" statement returns control to the first executable program line labeled as "loop" to continue the process. The "End" statement on the last line of the program terminates execution. In this example, the loop continues until power is removed. Although the End statement is never reached in this example, it is good programming practice to end every program file with an End statement.

9.4 Procedure for Programming a PIC with the MPlab IDE (Integrated Development Environment)

9.4.1 Introduction

The MPlab IDE is a development tool available from Microchip, Co. used to program their series of PIC microcontrollers. It contains all the tools necessary to write and compile code and to download hex files to the PIC processor. It also includes several debugging and simulation tools not used in this lab.

In this laboratory exercise you will create a project in the MPlab IDE. An MPlab IDE project includes several files including the source code, hex file, and project file. The source code contains the PicBasic code with the file extension ".bas". The hex file contains the code directly exportable to the PIC.

Programming PIC processors always requires three sequential steps. 1) Write or edit PicBasic code. 2) Compile this code to hexadecimal (hex). 3) Download the hex code to the PIC EEPROM. The PIC is then able to execute the code until it is erased or programmed again.

9.4.2 Procedures

Start MPLab by selecting:

Start button > Engineering Software pull-down menu > MPLab IDE

where the >'s indicate sequential submenus or dialog box items. If prompted to open a previous project, select NO.

Select:

Select Project > Install Language Tool...

Select all of the options as shown in Figure 9.2. The path to the Executable file PBP.EXE may be different on your computer. Click Ok when done.

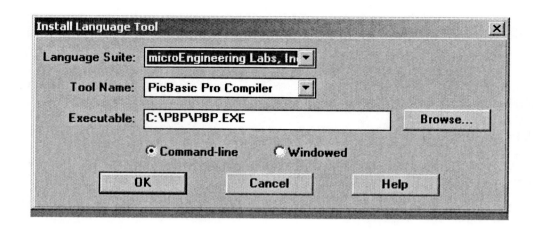

Figure 9.2 Language Tool Dialog Box

Then select:

Options > Development Mode...

As shown in Figure 9.3, make sure "None (Editor only)" and the correct processor are selected. Then click OK.

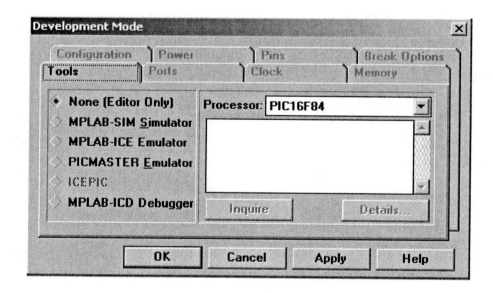

Figure 9.3 Development Mode Dialog Box

NOTE: the following paragraph is for reference only for future installation of software on new computers:

The following procedure has already been completed on Lab computers but will be required when installing the IDE on a new computer. For the PICBasic Pro compiler to work properly within the MPLAB IDE, the IDE needs to be able to locate file PM.EXE which is executed by the compiler PBP.EXE. The best way to enable this is to add the path of the PBP folder to the Windows PATH system environment variable. This is done by selecting: Start > Settings > Control Panel > System > Environment (in Windows NT) or Start > Control Panel > System > Advanced > Environment Variables (in Windows XP). Then double click on the PATH system variable. In the Value text box, position the cursor at the end of the existing string and add a semicolon (;) and the full path to the PBP folder. Then click Set and click OK. Then restart the computer.

NOTE: the following procedure is only necessary to allow proper PICSTART serial port communication for multiple processor PCs.

Open the Task Manager and click the Processes tab. Right click the NTVDM process then select "Set Affinity...". The "Processor Affinity" dialog should then appear. Uncheck all of the checked CPU's except CPU0 and click OK. Then close the task manager. This procedure only has to be done once per user session.

9.4.3   Starting a project

Select:

Click Project > New Project ...

and choose a folder in your own network file space and choose a meaningful name. This folder must exist already (i.e., you cannot create the folder from within the New Project dialog box). The extension ".pjt" is added to the filename automatically but may also be included at the end of the filename. **Choose a name with no spaces or symbols with no more than 8 letters.** Then click OK. An "Edit Project" dialog box will appear prompting for project properties. As shown in Figure 9.4, select microEngineering labs, Inc in the Language Tool Suite field. If a "Change Suite Warning" dialog box appears, click OK. Select yourfilename[.hex] in the Project Files list box and click "Node Properties..." As shown in Figure 9.5, deselect all options marked with a red check so that the command line field reads "-p16f84." Then click OK. Click OK again so that Edit Project dialog box closes.

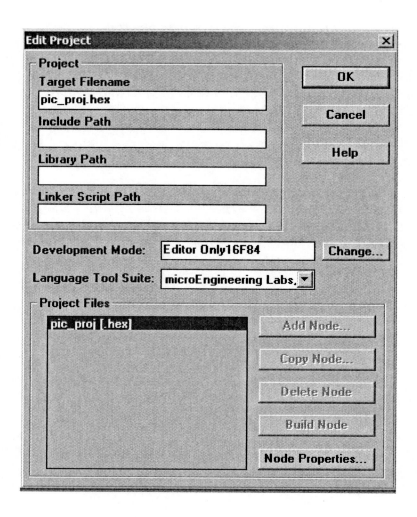

Figure 9.4 Edit Project Dialog Box

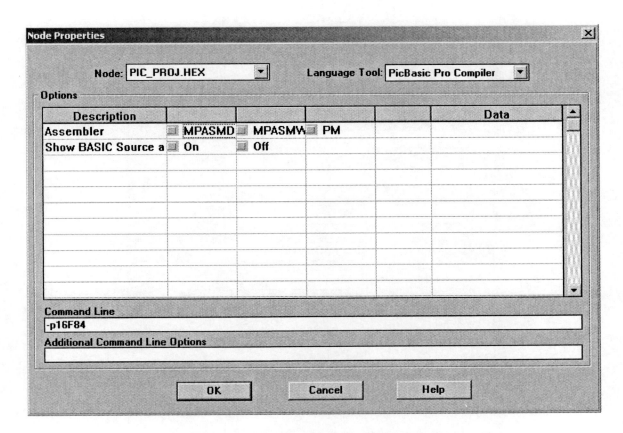

Figure 9.5 Node Properties Dialog Box

Start the source code file by selecting:

File > New…

After entering your PicBasic code choose Save As… and save the file as yourfilename.bas **in the same folder as your project file** from the previous step. **Make sure the file name includes the extension ".bas". It will not be added automatically.** Use the same file name that was used for the project. Only the extension differs. Click OK.

Add the source code file to the project by selecting:

Project > Edit Project…

Choose Add Node… and select the source code file from the previous step. As shown in Figure 9.6, this will append an indented yourfilename[.bas] node under yourfilename[.hex]. Click OK to close the Edit Project dialog box. To save your project files thus far, click File > Save All.

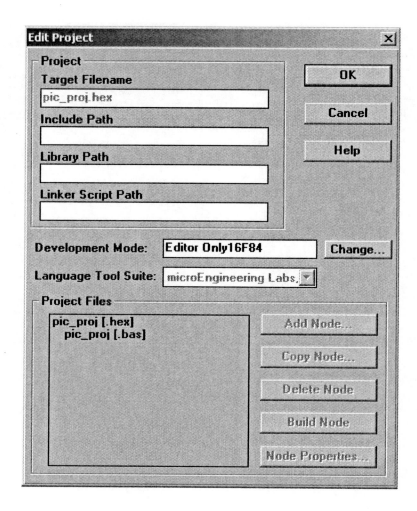

Figure 9.6 Edit Project Node Dialog Box

9.4.4   Adding line numbers to the editor window

Select:

Options > Current Editor Modes…

and click Show Line Numbers on the left hand side. This will help in locating errors when debugging your program. After compiling, MPlab IDE will display a text window showing any syntax errors by line number. Line numbers in the editor window will help locate those errors. Click OK.

9.4.5   Compiling your project

There are several ways to compile the project. Since this exercise does not require multinodal projects it is easiest to select the main toolbar icon that looks like a funnel (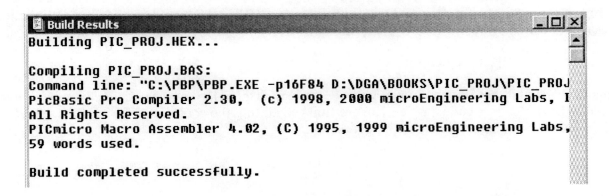). As the mouse cursor floats over this button, the status bar at the bottom of the window should read: "Make the Current Project." This means the source code will be converted to assembly code and then to hexadecimal code automatically. Click the button. It takes a few seconds to compile. As shown in Figure 9.7, a "Build Results" window will appear showing the results of the compilation. If compilation was successful, the message "Build completed successfully" will appear as the last line in the window. Otherwise, syntax errors in the code prevented compilation and generation of the hex file. The errors are listed by line number. The window also reports the amount of memory used by your program. In this case, 59 words out of a maximum of 1024 words (1K) are used. Any errors must be corrected and the program re-compiled. After a successful compilation, the IDE is ready to download the hex code to the PIC. When a project is compiled it is automatically saved. Close the Build Results window by clicking on the X in the upper right hand corner of the window.

```
Build Results                                              _ | □ | X |
Building PIC_PROJ.HEX...

Compiling PIC_PROJ.BAS:
Command line: "C:\PBP\PBP.EXE -p16F84 D:\DGA\BOOKS\PIC_PROJ\PIC_PROJ
PicBasic Pro Compiler 2.30,  (c) 1998, 2000 microEngineering Labs, I
All Rights Reserved.
PICmicro Macro Assembler 4.02, (C) 1995, 1999 microEngineering Labs,
59 words used.

Build completed successfully.
```

Figure 9.7 Build Results Report

9.4.6   Programming the PIC

Select:

PICSTART Plus > Enable Programmer

If the PICSTART Plus development programmer is connected correctly to the serial port and is plugged in, two or three windows will open. One will contain the contents of the most recently compiled hex file. The second window is the PICSTART Plus Devices Programmer interface. The third window is the Configuration Bits window. If this window is not visible, click the Configuration Bits button.

In the configuration bits window, make sure the XT oscillator is selected, the Watchdog timer is on (unless you are working with an assembly source file that uses this feature), the Power Up Timer is On, and the code protect option is off.

On the PicStart Plus Development Programmer (attached to the serial port of the computer), make sure the lock lever is pushed down. Then insert the PIC IC into the programmer, and lock it by pulling the lever up. Then click Program in the PicStart Plus Development Program window to initiate download of the hex file. The Program/Verify window will appear, the download will take a few seconds, and the window will disappear automatically when the download completes successfully. The PIC is now ready to be removed from the programmer and inserted into your application hardware. When power is applied, the program will begin to execute. If further software changes are needed, edit your source code and continue with Section 9.4.5.

When you are done, select File > Save All and close MPLAB. When prompted to save the project, click OK.

9.4.7   Editing your project

In the procedure outlined in Section 9.4.3 and 9.4.4 several steps were required to establish the project files and settings. When you edit your program at some later time, it is not necessary to repeat these steps. To open and edit a project, start MPLAB. If you do not want to open the previously edited project, click NO, and then select:

Project > Open Project ...

This will open all the files associated with the project, display the source code, and load the most recently compiled hex code. When you are finished editing the source code, click File > Save to save your work. Then continue with Section 9.4.5.

9.5   Using Interrupts

An interrupt is a specially designated input to a microcontroller that changes the sequence of execution of a program. When there is a change of state of one or more of the input pins designated as interrupts, the program pauses normal execution and jumps to separate code called an interrupt service routine. When the service routine terminates, normal program execution resumes with the statement following the point where the interrupt occurred. The interrupt service routine is identified by the PICBasic ON INTERRUPT GOTO command. An example program called onint.bas follows:

```
' onint.bas
' Example of use of and interrupt and interrupt handler
' This program turns on an LED and waits for an interrupt on PORTB.0.  When RB0 changes
'  state, the program turns the LED off for 0.5 seconds and then resumes normal execution.

led    var    PORTB.7         ' define variable led

    OPTION_REG = $7f     ' enable PORTB pull-ups and detect positive edges on interrupt
    On Interrupt Goto myint '  define interrupt service routine location
    INTCON = $90         ' enable interrupt on pin RB0

' Turn LED on and keep it on until there is an interrupt
loop:  High led
    Goto loop
```

```
' Interrupt handler
        Disable                 ' do not allow interrupts below this point
myint:  Low led                 ' if we get here, turn LED off
        Pause   500             ' wait 0.5 seconds
        INTCON.1 = 0            ' clear interrupt flag
        Resume                  ' return to main program
        Enable                  ' allow interrupts again

End                             ' end of program
```

The onint.bas program turns on an LED using PORTB.7 until an external interrupt occurs. A switch or button connected to pin 6 (PORTB.0) provides the source for the interrupt signal. When the signal transitions from low to high, the interrupt routine executes, causing the LED to turn off for half a second. Control then returns to the main loop causing the LED to turn back on again. More detail is provided in the following paragraphs.

NOTE: when using constants in a program, the dollar sign ($) prefix indicates a hexadecimal value percent sign (%) prefix indicates a binary value.

The first active line uses the keyword var to create the variable name led to denote the pin identifier PORTB.7. In the next line the OPTION_REG is set to $7f (or %01111111) to enable PORTB pull-ups and to configure the interrupt to be triggered when a positive edge occurs on pin RB0. When pull-ups are enabled, the PORTB inputs are held high until they are driven low by the external input circuit (e.g., a switch or button wired to pin RB0). The option register is defined in more detail below.

The label "myint" is defined as the location to which the program control jumps when an interrupt occurs. The value of the INTCON register is set to $90 (or %10010000) to properly enable interrupts. Setting the INTCON.7 bit to 0 globally allows all interrupts, and setting the INTCON.4 bit to 1 specifically enables the PORTB.0 interrupt. The INTCON register is described in more detail below.

The two lines starting with label "loop" cause the program to continually maintain the led pin (PORTB.7) high which keeps the LED on. The continuous cycle created by the "Goto loop" statement is called an infinite loop since it runs as long as no interrupt occurs. Note that an active statement (such as: "High led") MUST exist between the label and Goto of the loop for the interrupt to function because PICBASIC checks for interrupts only after a statement is completed.

The final section of the program contains the interrupt service routine. Disable must precede the label and Enable must follow the Resume to prevent further interrupts from occurring until control is returned to the main program. The interrupt routine executes when control of the program is directed to the beginning of this routine (labeled by "myint") when an interrupt occurs on PORTB.0 (pin 6). At the identifier label "myint" the statement Low led sets PORTB.7 (pin 13) to a digital low turning off the LED in the circuit. The Pause statement causes a 500 milliseconds (half a second) delay, during which the LED remains off. The next line sets the INTCON.1 bit to zero to clear the interrupt flag. The interrupt flag was set internally to 1 when the interrupt signal was received on PORTB.0, and this bit must be reset to zero before exiting the interrupt routine. At the end of the myint routine, control returns back to the main program loop where the interrupt occurred.

9.5.1   Registers Related to Interrupts

In order to detect interrupts, two specific registers on the PIC must be initialized correctly. These are the option register (OPTION_REG) and the interrupt control register (INTCON). The function of the individual bits within both registers are defined below.

The definition for each bit in the first register (OPTION_REG) follows. Recall that the least significant bit (LSB) is on the right, and is designated as bit zero (b_0), while the most significant bit (MSB) is on the left, and is designated as bit 7 (b_7).

$$OPTION\_REG = \%b_7b_6b_5b_4b_3b_2b_1b_0$$

bit 7:   RBPU: PORTB Pull-up Enable Bit
      1 = PORTB pull-ups are disabled
      0 = PORTB pull-ups are enabled (by individual port latch values)
bit 6:   Interrupt Edge Select Bit
      1 = Interrupt on rising edge of RB0/INT pin
      0 = Interrupt on falling edge of RB0/INT pin
bit 5:   T0CS: TMR0 Clock Source Select Bit
      1 = Transition on RA4/TOCK1 pin
      0 = Internal instruction cycle clock (CLKOUT)
bit 4:   T0SE: TMR0 Source Edge Select Bit
      1 = Increment on high-to-low transition on RA4/TOCK1 pin
      0 = Increment on low-to-high transition on RA4/TOCK1 pin
bit 3:   PSA: Prescaler Assignment Bit
      1 = Prescaler assigned to the Watchdog timer (WDT)
      0 = Prescaler assigned to TMR0
bits 2-0:PS2: PS0: Prescaler Rate Select Bits

Bit Value	TMR0 Rate	WDT Rate
000	1 : 2	1 : 1
001	1 : 4	1 : 2
010	1 : 8	1 : 4
011	1 : 16	1 : 8
100	1 : 32	1 : 16
101	1 : 64	1 : 32
110	1 : 128	1 : 64
111	1 : 256	1 : 128

In the onint.bas example above, OPTION_REG was set to $7f which is %01111111. Setting bit 7 low enables PORTB pull-ups and setting bit 6 high causes interrupts to occur on the positive edge of a signal on pin RB0. Bits 0 through 5 are only important when using special timers and are not used in this example.

116

The definition for each bit in the second register (INTCON) follows:

bit 7: GIE: Global Interrupt Enable Bit
        1 = Enables all unmasked interrupts
        0 = Disables all interrupts

bit 6: EEIE: EE Write Complete Interrupt Enable Bit
        1 = Enables the EE Write Complete interrupt
        0 = Disables the EE Write Complete interrupt

bit 5: T0IE: TMR0 Overflow Interrupt Enable Bit
        1 = Enables the TMR0 interrupt
        0 = Disables the TMR0 interrupt

bit 4: INTE: RB0/INT Interrupt Enable Bit
        1 = Enables the RB0/INT interrupt
        0 = Disables the RB0/INT interrupt

bit 3: RBIE: RB Port Change Interrupt Enable Bit (for pins RB4 through RB7)
        1 = Enables the RB Port Change interrupt
        0 = Disables the RB Port Change interrupt

bit 2: T0IF: TMR0 Overflow Interrupt Flag Bit
        1 = TMR0 has overflowed (must be cleared in software)
        0 = TMR0 did not overflow

bit 1: INTF: RB0/INT Interrupt Flag Bit
        1 = The RB0/INT interrupt occurred
        0 = The RB0/INT interrupt did not occur

bit 0: RBIF: RB Port Change Interrupt Flag Bit
        1 = When at least one of the RB7:RB4 pins changed state
           (must be cleared in software)
        0 = None of the RB7:RB4 pins have changed state

In the onint.bas example above, INTCON was set to $90 which is %10010000. For interrupts to be enabled, bit 7 must be set to 1. Bit 4 is set to 1 to check for interrupts on pin RB0. Bits 6, 5, 3, and 2 are for advanced features and are not used in this example. Bits 0 and 1 are used to indicated interrupt status during program execution.

If more than one interrupt signal were required, bit 3 would be set to 1 which would enable interrupts on pins RB4 through RB7. In that case, INTCON would be set to $88 (%10001000). To check for interrupts on RB0 and RB4-7, INTCON would be set to $98 (%10011000). PORTA has no interrupt capability, and PORTB has interrupt capability only on pins RB0 and pins RB4 through RB7.

9.6   Procedure

(1)   Use an ASCII editor (e.g., Windows Notepad or MS Word - Text Only) to create the program "blink.bas" listed in Section 9.3. Save the file in a folder in your network file space.

(2)   Follow the procedure in Section 9.4 to compile the "blink.bas" program into hexadecimal machine code ("blink.hex") and to load this code onto a PIC.

(3)   Assemble and test the circuit shown in Figure 9.1. When power is applied, the LED should immediately begin to blink on and off, cycling once each second.

(4)   Repeat steps (1) through (2) for the "onint.bas" program listed in Section 9.5. Before constructing the circuit for onint.bas, identify the additional components required in Figure 9.8. Indicate the necessary changes in the figure and check with your Teaching Assistant to verify that your changes are appropriate. The program should turn on an LED attached to PORTB.7. An interrupt on PORTB.0 should cause the LED to turn off for half a second, and then turn back on again. This should be signaled by an single pole, single throw (SPST) switch or a normally open (NO) button. You must make sure to wire the switch or button such that the ON state applies 5 Vdc to the pin, and the OFF state grounds the pin. The input should not be allowed to "float" in the OFF state (i.e., it must be grounded through a 1kΩ resistor). When you are sure you have all components wired properly, apply power to the circuit and test it to determine if it is working properly.

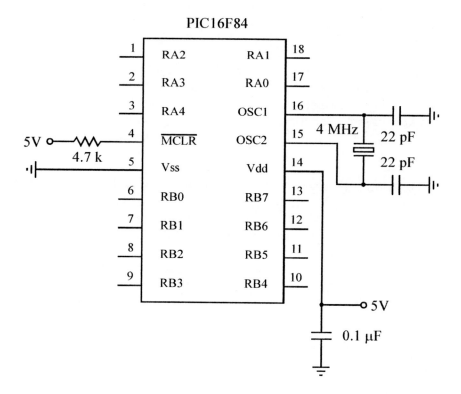

Figure 9.8 Circuit to be modified to run onint.bas

118

LAB 9 QUESTIONS

Group: _____   Names: _____

(1)   Explain all differences between PORTA and PORTB if using the pins for inputs.

(2)   For the onint.bas interrupt example, if the button is held down for more that 0.5 second and then released, is it possible that the LED would blink off again? If so, explain why. (Hint: consider switch bounce.)

(3)   Show two different ways to properly interface an LED to a PIC output pin. One circuit should light the LED only when the pin is high (this is called positive logic) and the other circuit should light the LED only when the pin is low (this is called negative logic).

(4)   Explain what you would observe if power were applied to a PIC loaded with the following
        code if an LED is connected to RB0 as shown in Figure 9.1.

```
before: High PORTB.0
        Pause 500
        Low PORTB.0
        Goto during
        Pause 100
        High PORTB.0
        Goto after
during: Low PORTB.0
        Pause 300
        High PORTB.0
        Pause 400
after:  Pause 200
        Low PORTB.0
        End
```

Laboratory 10

Programming a PIC Microcontroller - Part II

Required Components:
- 1 PIC16F84 (4MHz) or PIC16F84A (20MHz) 18P-DIP microcontroller
- 1 4MHz microprocessor crystal (20 pF)
- 2 22pF capacitors
- 1 0.1 μF capacitor
- 1 LED
- 1 470Ω resistor
- 1 4.7kΩ resistor
- 3 SPST microswitches or NO buttons
- 3 1kΩ resistors
- 1 7447 BDC to LED decoder
- 1 MAN 6910 or LTD-482EC seven-segment LED digital display
- 1 330Ω DIP resistor array

Required Special Equipment:
- Microchip PICSTART Plus development programmer
- Microchip's MPLAB integrated development environment windows software
- MicroEngineering Labs' PicBasic Pro compiler
- Demonstration hexadecimal counter circuit board containing a 555 timer circuit and a D flip-flop latch IC

10.1   Objective

This laboratory exercise builds upon the introduction to the PIC microcontroller started in the previous Lab. Here, input polling is introduced as an alternative to interrupts. You will learn how to configure and control the inputs and outputs of the PIC using the TRIS registers. You will also learn how to perform logic in your programs. You will first observe and describe the operation of a hexadecimal counter project provided by your TA. You will then create and test an alternative design using different hardware and software.

10.2   Hexadecimal Counter Using Polling

The first part of the laboratory involves the demonstration of an existing counting circuit using a PIC to activate the 7 segments of a digital LED display. At power-up a zero is displayed, and three separate buttons are used to increment by one, decrement by one, or reset the display to zero. The display is hexadecimal so the displayed count can vary from 0 to F. An input monitoring technique called polling is used in this example. With polling, the program includes a loop that continually checks the values of specific inputs. The output display is updated based on the values of these inputs. Polling is different from interrupts in that all processing takes place within the main program loop, and there is not a separate interrupt service routine. Polling has a disadvantage that if the program loop takes a long time to execute (e.g., if it performs complex control calculations), changes in the input values may be missed. The main advantage of polling is that it is very easy to program.

TRIS Registers

At power-up, all bits of PORTA and PORTB are initialized as inputs. However, in this example we require 7 output pins. Here, pins to be used as outputs are designated using special registers called the TRISA and TRISB. These registers let us define each individual bit of the PORT as an input or an output. In the previous examples that also required an output, this was not necessary since the High and Low commands set the TRIS registers automatically. In this example we will use assignment statements to set the PORT output values directly (e.g., PORTB = %00011001), requiring setting of the TRIS registers. Most PICBasic commands that use pins as outputs or inputs automatically set the TRIS register bits to appropriate values.

Setting a TRIS register bit to 0 designates an output and setting the bit to 1 designates an input. For example,

    TRISA = %00000000

designates all bits of PORTA as outputs and

    TRISB = %01110000

designates bits 4, 5, and 6 of PORTB as inputs and the others as outputs.

Note that since PORTA has only 5 usable bits (bits 0 through 4), the three most significant bits of PORTA are ignored and have no effect. At power-up all TRIS register bits are set to 1, so all pins are treated as inputs by default (i.e., TRISA=$FF and TRISB=$FF).

The code "counter.bas" for the up/down hex counter in listed below:

```
' counter.bas
' PicBasic hex up/down counter


' Declare variables
pins        var     byte[16]    ' an array of 16 bytes used to store the 7-segment display codes
I           var     byte        ' counter variable


' Initialize I/O pins
TRISA = %00000000           ' all PORTA pins initialized as outputs
                            '   (although only pins 0, 1, and 2 are used)
TRISB = %01110000           ' PORTB.4,5,6 pins initialized as inputs
                            '   (RB4: reset, RB5: increment, RB6: decrement)
                            ' all other PORTB pins initialized as outputs
```

' Initialize the pin values for the 7-segment LED digit display with segments as illustrated below
'
'
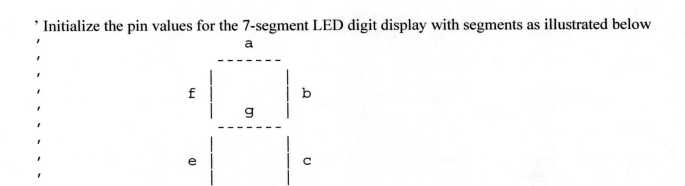
'
'
'
'
'
'
'
'
'
'

' Three of the pins in PORTA and four of the pins in PORTB are used as the seven outputs.
' The LED segments are assigned to the PORT bits as shown below:
'   PORTA   PORTB
' pins: 76543210   76543210
' segments: -----cde   ----bagf

' NOTE: 0 turns a segment ON and 1 turns it OFF since the PIC sinks current from the LED display
'   binary   hex   display
pins[0] = %00000010   ' 02   0
pins[1] = %00110111   ' 37   1
pins[2] = %01000001   ' 41   2
pins[3] = %00010001   ' 11   3
pins[4] = %00110100   ' 34   4
pins[5] = %00011000   ' 18   5
pins[6] = %00001000   ' 08   6
pins[7] = %00110011   ' 33   7
pins[8] = %00000000   ' 00   8
pins[9] = %00110000   ' 30   9
pins[10] = %00100000   ' 20   A
pins[11] = %00001100   ' 0C   b
pins[12] = %01001010   ' 4A   C
pins[13] = %00000101   ' 05   d
pins[14] = %01001000   ' 48   E
pins[15] = %01101000   ' 68   F
'   %0cdebagf   (correspondence between LED segments and pins[] bits)

' Initialize the display to zero
I = 0
Gosub Updatepins

123

```
' Main loop
loop:


  If (PORTB.4 == 1) Then     ' reset
    I = 0
    Gosub Updatepins
    Pause 100    ' 0.1 sec delay
  Endif

  If (PORTB.5 == 1) Then    ' increment
    If (I == 15) Then
      I = 0
    Else
      I = I + 1
    Endif

    Gosub Updatepins
    Pause 100    ' 0.1 sec delay
  Endif

  If (PORTB.6 == 1) Then    ' decrement
    If (I == 0) Then
      I = 15
    Else
      I = I - 1
    Endif

    Gosub Updatepins
    Pause 100    ' 0.1 sec delay
  Endif

Goto Loop       ' go back to the beginning of the loop and continue to poll the inputs


' Updatepins Subroutine
'   sends new output values to pins

Updatepins:
    ' Use the right shift operator to move the top MSB's of pins[I] to the 4 LSB's of PORTA
    '    padding the 4 MSBs of PORTA with 0's
    PORTA = pins[I] >> 4
    ' Use logic to retain the 4 MSB's of PORTB and replace the 4 LSB's of PORTB by
    '    by the 4 LSB's of pins[I]
    PORTB = (PORTB | %00001111) & (pins[I] | %11110000)
    Return

End                ' End of program
```

124

After the initial comments labelling the program, the pins variable is defined as an array of 16 bytes. Elements in the array are accessed by the syntax pins[I], where I is the index having values from 0 through 15. Binary values for the pins array elements are set in the "Initialize the pin values ..." section. The mapping of these bits to individual segments on the 7-segment counter display, and the structure of the counter circuit dictates the assignment of each of these bits to a specific segment. The 8 bits in each byte are grouped into 2 sets of 4 bits: the left 4 bits (most significant bits: MSB's) assign values to the pins set by PORTA, and the right 4 bits (least significant bits: LSB's) assign values to the pins set by PORTB. Again, these depend on the function of the circuit attached to the PIC. Bits 4, 5, and 6 of the pins[I] variable are output through PORTA pins to segments e, d, and c of the display, respectively. Bits 0, 1, 2, and 3 of the pins[I] variable are output through PORTB pins to segments f, g, a, and b of the display, respectively. Note that bit 7 is set to 0 for each element in the pins[I] variable. Also, note that bit values assume negative logic where a 0 turns the segment on and a 1 turns the segment off.

The TRIS registers are set to determine the I/O status of the pins in PORTA and PORTB. Since all bits in TRISA are 0, all pins corresponding to PORTA are set as outputs. Note that PORTA bits 5, 6, and 7 have no function since no pins actually exist on the PIC to correspond to these values. The TRISB register value is set so that PORTB bits 4, 5, and 6 are inputs (each of these 3 bits is set to 1), while the other 5 pins of PORTB are set as outputs. Each of these three input pins for PORTB is attached to a separate button. Depending upon which button is pressed, the counter will either increment by one, decrement by one, or reset to zero using the hexadecimal counting sequence.

The polling loop used to check for button input is in the "Main loop" of the program. The first IF statement checks whether the button attached to PORTB.4 is down. If it is, the index for the pins[I] variable is set to 0 so that the display will be zero. The Updatepins routine is called to update the value displayed as described in detail below. Then a pause occurs for 100 milliseconds (0.1 sec). The polling then continues by checking PORTB.5, and if the value is high, then the hexadecimal count is incremented by incrementing the index for the pins[I] variable. The internally nested IF statement checks if the index exceeds the allowed value of 15, and if it does the index is reset to 0. The display effectively will count from 0 through 15 as the button is repeatedly depressed or held down, but will cycle back to 0 after an F has been displayed (the highest digit value in hexadecimal). Again a call to Updatepins updates the display, and a 0.1 second pause occurs. The pause prevents the count from updating too quickly while the button is being held down before it is released. If the button is held down for more than 0.1 second the counter will increment every 0.1 second. Then PORTB.6 is checked and a value of 1 will cause a decrement in the index for the pins[I] variable. Once the display reaches a minimum value of zero, the routine will cycle back to F for the next hexadecimal value.

The most direct action in displaying the output occurs in the Updatepins routine. A simple assignment statement (a statement containing an equal sign =) performs the write to the pins that change the segments in the display. Recall that only three of the five available output pins are used on PORTA (the LSB's of PORTA). Since the pins[I] variable stores values for PORTA in bits 4 through 7 (the MSB's of pins[I]), these bits must be extracted and shifted. The right shift operator (>>) is used to shift the four MSB's four places to the right to become the four LSB bits 0 through 3. The four MSB's are replaced with 0's as a result of the shift. The result is written to PORTA by the assignment statement. Output to PORTB is more complex since the procedure seeks to maintain the existing values for bits 4 through 7 (the MSB's of PORTB) since these bits are reserved for the button inputs, while changing bits 0 through 3 (the LSB's of PORTB). Boolean logic operators for OR (|) and AND (&) are used to carry out this process. The OR in the left set

125

of parentheses maintains the four MSB's of PORTB, and sets the four LSB's to 1. The OR in the right set of parentheses maintains the four LSB's of the pins[I] variable, and sets the four MSB's to 1. When the two results are ANDed, the four MSB's of PORTB are maintained, while the four LSB's are changed to the values found in the currently indexed pins[I] element. The assignment to PORTB effectively writes the LSB's to the pins, which turns the respective segments on or off. The display is updated every time a call is made to the Updatepins routine.

As shown in Figure 10.1, latching of button values is accomplished by the use of D flip-flops (74LS175) in the circuit. Also key to the circuit, but not shown in the figure, is the use of a 555 timer circuit to create a 100 Hz clock signal. On each positive edge of clock pulse, the current states of the buttons are stored (latched) in the D flip-flips on the 74LS175 IC. Together, the timer and flip-flops perform a hardware debounce. The latched values are read by the PIC each time the program passes through the polling loop. Note that the buttons are shown wired in the figure with negative logic, where the button signal is normally high and goes low when it is pressed. The software above assumes positive logic instead where the button signal is normally low and goes high when pressed. This is easily accomplished by using a pull-up resistor from ground instead of a pull-down resistor from 5V.

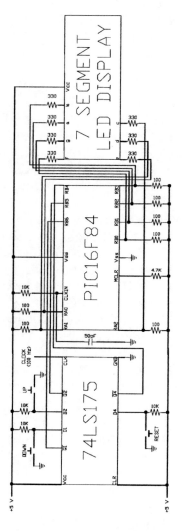

Figure 10.1 Circuit Diagram for Hexadecimal Counter

10.3   An Alternative Design

The hardware and software design above can be simplified if you use PORTA for the three button inputs and PORTB for all seven of the LED segment outputs. This was not done for the example above since the hardware was originally designed and built by a graduate student assuming that interrupts would be used. Interrupts are available only in PORTB, and if the three buttons were attached to PORTB, only five bits in PORTB would be available to be used as outputs. Furthermore, PORTA only has five bits available. Because we need seven bits to drive the display, both PORTA and PORTB were used for the seven outputs.

An alternative design is outlined below using PORTB for all seven outputs and PORTA for the three inputs. This dramatically simplifies the Updatepins subroutine obviating the need for the complex logic manipulations of the bits. The changes required to the hardware and software for the alternative design follow.

The pins in PORTB are assigned and connected to the LED segments as follows:

```
pins:        %76543210
segments:    %-cdebagf
```

The TRIS registers are initialized as follows:

```
TRISA = %00001110          ' PORTA.1,2,3 pins are inputs
TRISB = %00000000          ' all PORTB pins are outputs (although, pin 7 is not used)
```

Then the simpler Updatepins subroutine is:

```
Updatepins:
  PORTB = pins[I]
  Return
```

where the bit values in the pins[I] array element are written directly to the PORTB bits driving the LED segments.

Switch debounce can be performed in software instead of hardware eliminating the need for the 555 timer and D flips-flop portions of the circuit. Figure 10.2 shows the hardware for the alternative design (using bits 1,2, and 3 on PORTA).

127

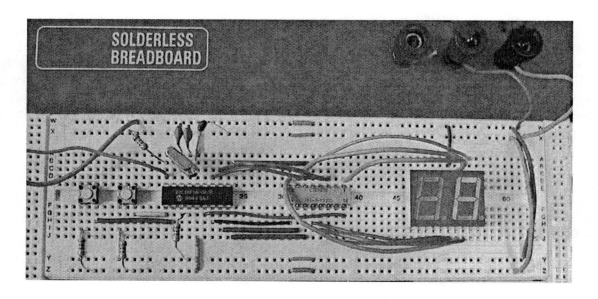

Figure 10.2 Alternative Design Hexadecimal Counter Circuit

The switch bounce that can occur when the button is pressed is not a problem in the original software presented above because a 0.1 sec pause gives the button signals more than enough time to settle. However, if a button if held down for more than 0.1 sec and then released, any bouncing that occurs upon release could cause additional increments or decrements. One approach to perform debouncing for the button release is to use a delay in software that waits for the bounce to settle before continuing with the remainder of the program. Here is how the code could be changed for the increment button:

```
' Continue to increment every 0.1 sec while the increment button is being held down

While (PORTA.1 == 1)
        If (I == 15) Then
                I = 0
        Else
                I = I + 1
        Endif

        ' Update the display
        Gosub Updatepins

        ' Hold the current count on the display for 0.1 sec before continuing
        Pause 100
Wend


' Pause for 0.01 sec to allow any switch bounce to settle after button release
Pause 10
```

The decrement button would be handled in a similar fashion. No debounce is required for the reset button because multiple resets in a short period of time (e.g., the few thousandths of a second when bouncing occurs) do not result in undesirable behavior.

Yet another alternative design that would further simplify the software would be to use a 7447 IC for BCD-to-7-segment decoding. This would eliminate the need for the pins array that does the decoding in software, but it would add an additional IC to the hardware design. Also, the 7447 displays non-alphanumeric symbols for digits above 9, instead of the hexadecimal characters (A, b, C, d, E) that we control with the pins array.

10.4 Procedure

(1) Observe the demonstration of the hexadecimal counter circuit. Study the program listed in Section 10.2 and fully test all functionality of the circuit. Examine the effect of holding down a button. Examine the effects of holding down multiple buttons at once.

(2) Draw a complete and detailed schematic required to implement the alternative counter design described in Section 10.3. Figure 10.3 shows useful information from the MAN6910 datasheet. Have your TA check your diagram before you continue.

(3) Use an ASCII editor (e.g., Windows Notepad or MS Word - Text Only) to create the program necessary to control the alternative design. Name it "counter.bas". Save the file in a folder in your network file space named "counter."

(4) Follow the procedure in the previous laboratory exercise to compile the program and load it onto a PIC.

(5) Assemble and fully test your circuit with the programmed PIC. Have your TA verify that it works properly.

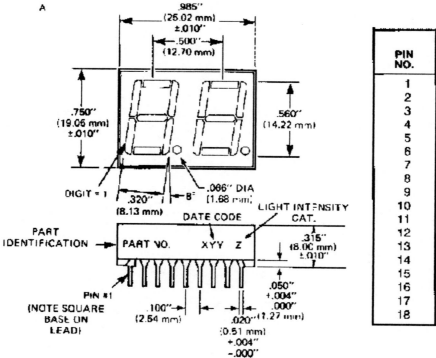

PIN NO.	A MAN6910
1	Cathode E 1
2	Cathode D 1
3	Cathode C 1
4	Cathode D.P. 1
5	Cathode E 2
6	Cathode D 2
7	Cathode G 2
8	Cathode C 2
9	Cathode D.P. 2
10	Cathode B 2
11	Cathode A 2
12	Cathode F 2
13	Anode Digit 2
14	Anode Digit 1
15	Cathode B 1
16	Cathode A 1
17	Cathode G 1
18	Cathode F 1

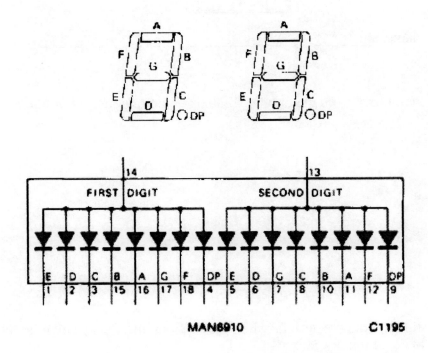

Figure 10.3 MAN6910 Datasheet Information

LAB 10 QUESTIONS

Group: _____ Names: _____

(1)   Rewrite "onint.bas" from the previous Lab using polling instead of interrupts.

(1)   For the original counter design, when the 'up' button is held down awhile the PIC will continue to count up. Explain why.

(2)   For the original counter design, explain what happens when the 'up' and 'down' buttons are held down together. Why does this happen?

(3)   For the alternative counter design, why is the 555 and D flip-flop hardware no longer required?

(4)   Explain how switch bounce could have a negative impact with the alternative design if the 0.01 sec software pause were not included.

(5)   Explain why debounce software is not required for the reset button in the alternative design.

(6)   For the demonstrated counter, how would you create the functionality in the Updatepins subroutine using individual bit references (e.g. PORTA.0 = pins[I].0)?

Laboratory 11

Pulse Width Modulation Motor Speed Control with a PIC

Required Components:
- 1 PIC16F84 (4MHz) or PIC16F84A (20MHz) 18P-DIP microcontroller
- 1 4MHz microprocessor crystal (20 pF)
- 2 22pF capacitors
- 1 0.1 μF capacitor
- 1 12-button numeric keypad
- 1 NO pushbutton switch
- 1 Radio Shack 1.5-3V DC motor (RS part number: 273-223) or equivalent
- 1 IRF620 power MOSFET
- 1 flyback diode (e.g., the IN4001 power diode)
- 3 1kΩ resistors
- 1 2kΩ resistor (or 2 1kΩ resistors)
- 1 3kΩ resistor (or 3 1kΩ resistors)
- 3 red LEDs
- 1 green LED
- 4 330Ω resistors or a 330Ω 8-resistor DIP

Required Special Equipment:
- Microchip PICSTART Plus development programmer
- Microchip's MPLAB integrated development environment windows software
- MicroEngineering Labs' PicBasic Pro compiler

11.1   Objective

The objective of this laboratory exercise is to design and build hardware and software to implement pulse-width modulation (PWM) speed control for a small permanent-magnet dc motor. You will also learn how to interface a microcontroller to a numeric keypad and how to provide a numerical display using a set of LEDs.

11.2   Introduction

Pulse Width Modulation

Pulse width modulation (PWM) offers a very simple way to control the speed of a dc motor. Figure 11.1 illustrates the principles of operation of PWM control. A dc voltage is rapidly switched at a fixed frequency f between two values ("ON" and "OFF"). A pulse of duration t occurs during a fixed period T, where

$$T = \frac{1}{f} \tag{11.1}$$

The resulting asymmetric waveform has a ***duty cycle*** defined as the ratio between the ON time and the period of the waveform, usually specified as a percentage:

$$\text{duty cycle} = \frac{t}{T}100\% \tag{11.2}$$

As the duty cycle is changed (by varying the pulse width t), the average current through the motor will change, causing changes in speed and torque at the output. It is primarily the duty cycle, and not the value of the power supply voltage, that is used to control the speed of the motor.

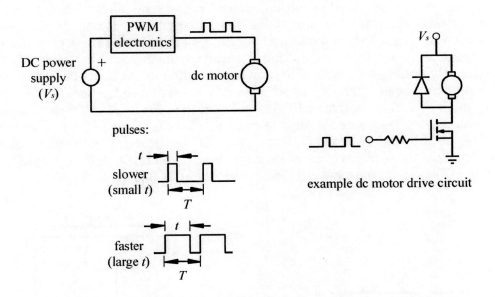

Figure 11.1 Pulse-width Modulation (PWM)

With a PWM motor controller, the motor armature voltage switches rapidly, and the current through the motor is affected by the motor inductance and resistance. For a fast switching speed (i.e., large f), the resulting current through the motor will have only a small fluctuation around an average value, as illustrated in Figure 11.2. As the duty cycle gets larger, the average current gets larger and the motor speed increases.

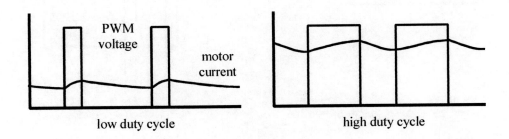

Figure 11.2 PWM voltage and motor current

The type of PWM control described here is called "open loop" because there is no sensor feedback for speed. This results in a simple and inexpensive design, but it is not possible to achieve accurate speed control without feedback. For precision applications (e.g., industrial robotics), a speed sensor (e.g., a tachometer) is required to provide feedback to the electronics or software in order to adjust the PWM signal in real-time to maintain the desired speed. See Section 10.5.3 in the textbook for more information.

<u>Numeric Keypad Interface</u>

Figure 11.3 illustrates the appearance and electrical schematic for a common 12-key **numeric keypad**. When interfaced to a microcontroller, a keypad allows a user to input numeric data. A keypad can also be used simply as a set of general-purpose normally-open (NO) pushbutton switches. The standard method to interface a keypad to a microcontroller is to attach the four row pins to inputs of the microcontroller and attach the three column pins to outputs of the microcontroller. By polling the states of the row inputs while individually changing the states on the column outputs, you can determine which button is pressed. See Section 7.7.1 in the textbook for more information. An alternative method to interface the keypad, if you do not have the luxury of seven spare I/O lines, is to wire the keypad through a set of resistors in series with a capacitor to ground. This allows you to use the PicBasic Pro "Pot" command to determine which button is pressed by reading the effective resistance of the keypad through a single pin of the microcontroller. The circuit presented in the next section uses this method.

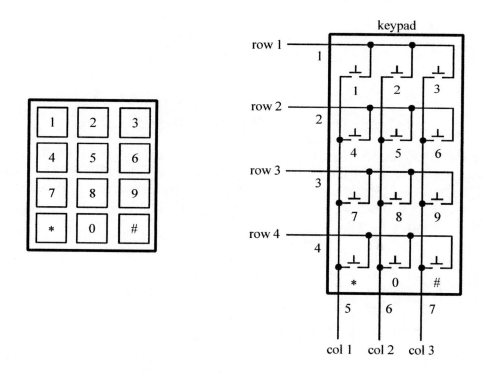

a) device appearance   b) device electrical schematic

<u>Figure 11.3</u> Standard 12-key Numeric Keypad

136

11.3   Hardware and Software Design

The hardware and software required for this exercise will be designed using the microcontroller-based design procedure presented in Section 7.9 of the textbook. Each step is presented below.

(1)   *Define the Problem.*

Use a PIC16F84 microcontroller to design a pulse-width modulation speed controller for a small permanent magnet dc motor. The user should be able to change the speed via three buttons of a standard 12-key numeric keypad. One button (the 1-key) should increase the speed setting, a second button (the 4-key) should decrease the speed setting, and the third button (the *-key) should start the motor at the selected speed. The speed setting should be displayed graphically via a set of 4 LEDs. The speed setting should vary from "slow" to "fast" according to a scaled number ranging from 0 to 15 so the full range can be depicted on the LED display. The motor should run at a constant speed until the motion is interrupted by the user with the press of a pushbutton switch. All inputs and outputs for this problem are digital and they are as follows:

inputs:

- 3 buttons on the numeric keypad to increase and decrease the speed and to start the motion.

- 1 pushbutton switch to interrupt the constant speed motor motion.

outputs:

- 4 LEDs to indicate a relative speed setting from "slow" (0) to "fast" (15) as a binary number.

- 1 pulse-width modulation (on-off) signal for the motor.

(2)   *Select an appropriate microcontroller.*

For this problem, we will use the PIC16F84 whose 13 lines of digital I/O provide more than enough capability for our I/O requirements.

(3)   *Identify necessary interface circuits.*

To help you learn how to use a numeric keypad in the most efficient way, we will show you how to connect the rows and columns of the keypad through a network of resistors in series with a capacitor through a single pin on the PIC. With the help of the PICBasic Pro command "Pot," we can determine which button is pressed based on the time constant of the resulting RC network. The resistance will change based on which button is pressed. Only a single digital input is required to implement this method.

The motor speed will be controlled with a pulse-width modulation signal. We will use a power MOSFET to switch current to the motor. The gate of the MOSFET will be connected directly to a digital output pin on the PIC. The motor is placed on the drain side of the MOSFET with a diode for flyback protection. A MOSFET is easier to use than BJT because it does not require a base (gate) resistor, and you need not be concerned with base current and voltage biasing.

The LEDs will be connected directly to four digital outputs through current-limiting resistors to ground. When the output goes high, the LED will turn on.

(4)   *Decide on a programming language.*

For this laboratory exercise, we will use PicBasic Pro.

(5)   *Draw the schematic.*

Figure 11.4 shows the complete schematic showing all components and connections. Figure 11.5 shows a photograph of a completed design. The keypad is attached to PORTA.2 and the stop button is attached to PORTA.3. The keypad is wired such that different resistors are in series with a fixed capacitor depending upon which button is held down (1kΩ for the 1-key, 2kΩ for the 4-key, and 3kΩ for the *-key). The LEDs are attached to the four lowest order bits of PORTB. This allows the speed setting (0 to 15) to be output to PORTB directly (e.g., PORTB = speed). The result is a binary number display of the current speed where the green LED represents the LSB. The motor PWM signal is on PORTA.1.

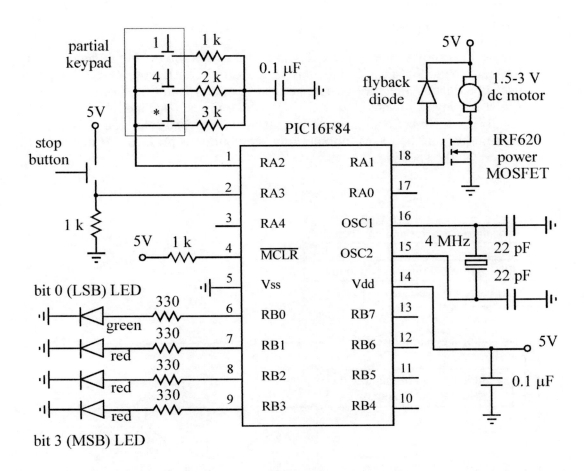

Figure 11.4 Complete Schematic Showing All Components and Connections

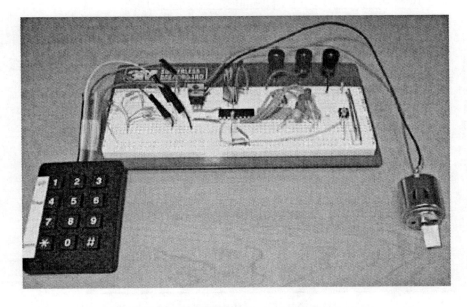

Figure 11.5 Photograph of Actual Design

(6) *Draw a program flowchart.*

Figure 11.6 shows the complete flowchart for this problem with all required logic and looping. Note that the LED display is active only during the keypad loop while the user is adjusting the speed. The keypad is polled using the Pot command and the speed display is updated approximately three times a second. Each keypad button results in a different resistance value that can vary over a small range. The motor runs continuously in the PWM loop until the stop button is pressed. At that point the user can adjust the speed again.

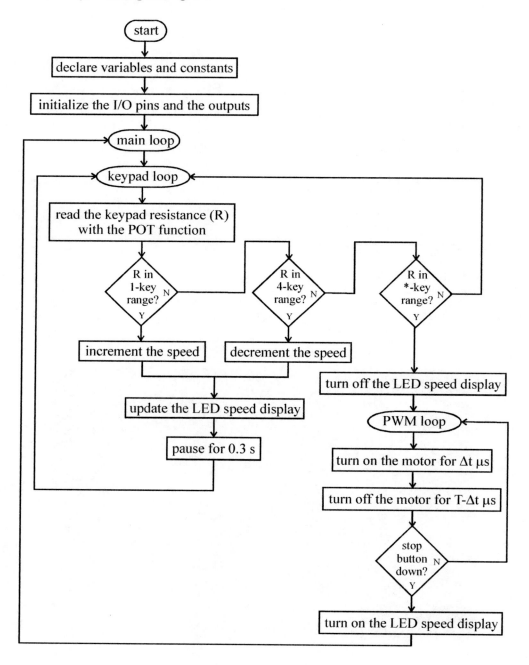

Figure 11.6 Complete Program Flowchart

The PicBasic Pro code ("PWM.bas") corresponding to the flowchart shown in Figure 11.6 using the hardware illustrated in Figure 11.4 follows. The code is commented throughout with remarks so it should be self-explanatory. Whenever you write programs, you should always include copious remarks so you and others (e.g., co-workers and bosses) can later interpret what you have done.

```
' PWM.bas
'
' Controls the speed of a DC motor using pulse-width modulation (PWM).  The speed is adjusted
' via user input with three buttons (increase, decrease, and enter) on a numeric keypad.  The relative
' speed is stored as a number that ranges from 0 (corresponding to 60% duty cycle) to 15
' (corresponding to 80% duty cycle).  The current value of the speed is displayed graphically
' with a set of 4 LEDs that show the bits of the equivalent binary number.


' Define pin assignments, variables, and constants
led0            Var     PORTB.0         ' LSB (bit 0) green LED
led1            Var     PORTB.1         ' bit 1 red LED
led2            Var     PORTB.2         ' bit 2 red LED
led3            Var     PORTB.3         ' MSB (bit 3) red LED

motor           Var     PORTA.1         ' PWM output pin to motor MOSFET gate
change          Var     PORTA.3         ' button causing the motor to stop for
                                        '   speed adjustment
speed           Var     BYTE            ' User-input speed
MAX_SPEEDCon    15                      ' Maximum relative speed
T               Var     WORD            ' pulse period in milliseconds
t               Var     WORD            ' pulse width (high state)
T_t             Var     WORD            ' pulse down (low state) time: (T - t)

pot_pin         Var     PORTA.2         ' keypad pin for POT command
SCALE           Con     255             ' Pot statement scale factor
pot_val         Var     BYTE            ' value returned by POT command


' Initialize the I/O pins
TRISA = %11101                  ' designate PORTA pins as inputs and output (RA1)
TRISB = %00000000               ' designate PORTB pins as outputs
PORTB = 0
Low motor                       ' make sure the motor remains off initially

' Initialize the speed display information
T = 30000                       ' pulse period in microseconds
speed = 7                       ' select a medium speed to begin (the middle of the 0 to 15 range)
PORTB = speed                   ' display the speed as a binary number on the 4 LEDs
```

```
' Main Loop
loop:
        ' User speed change loop
        While (1)       ' until the * key on the keypad is pressed
                ' Read the keypad resistance
                POT pot_pin, SCALE, pot_val

                ' Check for the 1-key to increase the speed
                If (pot_val > 30) && (pot_val < 95) && (speed < MAX_SPEED) Then
                        speed = speed + 1
                        PORTB = speed
                        Pause 300
                Else
                  ' Check for the 4-key to decrease the speed
                  If (pot_val > 95) && (pot_val < 160) && (speed > 0) Then
                        speed = speed - 1
                        PORTB = speed
                        Pause 300
                Else
                        ' Check for the *-key to start motor motion
                        If (pot_val > 160) Then run
                Endif
                Endif
        Wend

' Run the motor until the user presses the stop button
run:
        ' Turn off the LEDs
        PORTB = 0

        ' Initialize the pulse information
        t = T/5 / MAX_SPEED * speed + T/5*3         ' duty cycle range = 60% to 80%
        T_t = T - t

        ' Run the PWM until the user presses the stop button
        While (change = 0)
                High motor
                Pauseus t
                Low motor
                Pauseus T_t
        Wend

        ' Turn the LED speed display back on
        PORTB = speed
Goto loop


' End of the program (never reached)
End
```

The variable "speed" stores a relative measure of the motor speed as an integer that varies from 0 (slow) to 15 (fast). A speed of 0 corresponds to a duty cycle of 60% and the a speed of 15 corresponds to a duty cycle of 80%. These duty cycle percentages were determined experimentally to produce a good range of motor speeds using a 5 V supply. (Note - the motor is rated at only 1.5 to 3 V so duty cycles over 80% would result in excessive speed that could damage the motor.)

One not so obvious challenge in the program is how the variable "t" is calculated. Because PicBasic Pro stores variables and does arithmetic with limited size integers, you have to be careful with truncation and overflow effects when performing calculations. For example, the equation:

$$t = T/5 \ / \ MAX\_SPEED * speed + T/5*3 \qquad\qquad (11.3)$$

would not work properly if it were written as:

$$t = speed \ / \ MAX\_SPEED * T/5 + T/5*3 \qquad\qquad (11.4)$$

or as:

$$t = T/5 * speed \ / \ MAX\_SPEED + T/5*3 \qquad\qquad (11.5)$$

The variable speed can vary from 0 to 15, so from Equation 11.3 where MAX_SPEED is 15, t can vary from 3/5 T (60% of T) to 4/5 T (1/5 T + 3/5 T = 80% of T). Note that parentheses are not required to have the calculations in the equation execute in the correct order because, as with all programming languages, PicBasic Pro gives higher precedence to multiplication and division (which occur from left to right), than with addition and subtraction. Therefore, to PicBasic Pro, Equation 11.3 looks like:

$$t = (((T/5) \ / \ MAX\_SPEED) * speed) \ + \ ((T/5) * 3) \qquad\qquad (11.6)$$

There is a problem with Equation 11.4 due to integer arithmetic **truncation**. Because "speed" varies from 0 to 15 and MAX_SPEED is 15, for all values of speed except 15 (0 through 14), the integer fraction "speed/MAX_SPEED" will be truncated to 0 (because the result of the division is less than 1) before the remaining calculations are executed. Equation 11.5 will not work as desired because, for high speed values the product "(T/5)*speed" will exceed the largest value that can be stored with a 16-bit WORD variable ($2^{16} - 1 = 65,535$). This is called **overflow**. For all values of "speed" greater than 10, the product "(T/5)*speed" will be truncated to 65,535, throwing off the remaining calculations. In Equation 11.3, the order of calculations is chosen carefully so no truncation or overflow occurs.

The If statements in the While loop check to determine the range within which the Pot command variable "pot_val" falls. This allows the program to determine which button on the keypad is pressed. A separate calibration program was written to determine the appropriate values for the range limits. This program ("PWM_cal.bas" below) uses the same hardware as for the program above ("PWM.bas"), but here the LEDs are being used to graphically display the value returned by the Pot command. The three red LEDs blink individually and sequentially to indicate the number of 100s, 10s, and 1s in the "pot_val" number. The green LED is flashed as a signal between each red LED's digit value display. If you had an liquid crystal display (LCD) in your design, it would be a simple matter to display the decimal number on the LCD for easy viewing. However, to use an LCD with the Pic Basic Pro command "Lcdout" requires 7 I/O pins, and many project designs will not have enough spare pins to drive the display. If you only have one or a few output pins available, blinking LEDs offer an alternative method to graphically display the values of numbers within your running program. In "PWM_cal.bas," since we have four LEDs, we used three different LEDs to indicate the different decimal places for the number. If you didn't have multiple LEDs in your design or if you only had one pin to spare, you could achieve the same result by blinking a single LED with pauses between each digit number display.

Through testing with the "PWM_cal.bas" program, using a "Pot" command scale value of 255, we found the following values for the three keys: 65 for the 1-key, 128 for the 4-key, and 189 for the *-key. That is why the following pot_val ranges where used in the "PWM.bas" program: 30 to 95 for the 1-key, 95 to 160 for the 4-key, and above 160 for the *-key. The nominal values (65, 128, and 189) fall in the middle of these ranges allowing for small random fluctuations due to temperature and connection resistance changes. Refer to the PicBasic Pro manual for details on how to select an appropriate value for the "Pot" command scale value. The value 255 is appropriate for the resistance and capacitance values we selected.

' PWM_cal.bas

```
' Displays the Pot values for the keypad buttons by blinking the upper three red LEDs.  Each
' LED is blinked individually to indicate the number of 100s, 10s, and 1s in the
' Pot value number.  The green LED is flashed once between each blinking red LED display.

' Define variables, pin assignments, and constants
led0            Var     PORTB.0         ' LSB (bit 0) LED
led1            Var     PORTB.1         ' bit 1 LED
led2            Var     PORTB.2         ' bit 2 LED
led3            Var     PORTB.3         ' MSB (bit 3) LED

motor           Var     PORTA.1         ' PWM output pin to motor MOSFET gate

pot_pin         Var     PORTA.2         ' keypad pin for POT command
SCALE           Con     255             ' Pot statement scale factor
pot_val         Var     BYTE            ' value returned by POT command

i               Var     BYTE            ' loop variable
digs            Var     BYTE            ' digit number for each decimal place


' Initialize the I/O pins
TRISA = %11101                  ' designate PORTA pins as inputs and output (RA1)
TRISB = %00000000               ' designate PORTB pins as outputs
PORTB = 0
Low motor                       ' make sure the motor remains off


' User speed change loop
enter:
        POT pot_pin, SCALE, pot_val

        ' Flash the LSB green LED and blink each of the upper 3 red LEDs to indicate the number of
```

```
'   100s, 10s, and 1s in pot_val
PORTB = 0

High led0
Pause 500
Low led0
Pause 100
digs = pot_val / 100
For i = 1 To digs
        High led3
        Pause 300
        Low led3
        Pause 300
Next i

pot_val = pot_val - digs*100
High led0
Pause 500
Low led0
Pause 100
digs = pot_val / 10
For i = 1 To digs
        High led2
        Pause 300
        Low led2
        Pause 300
Next i

digs = pot_val - digs*10
High led0
Pause 500
Low led0
Pause 100
For i = 1 To digs
        High led1
        Pause 300
        Low led1
        Pause 300
Next i
Goto enter

' End of program (never reached)
End
```

(8)   *Build and test the system.*

That is your job using the procedure below.

11.4 Procedure

(1)　Use an ASCII editor (e.g., Windows Notepad or MS Word - Text Only) to create the program "PWM_cal.bas" listed in Section 11.3. Save the file in a folder in your network file space.

(2)　Follow the procedure in Section 9.4 of Lab 9 to store your program in a PIC microcontroller that you can insert into your circuit.

(3)　Build the circuit shown in Figure 11.4 and insert the PIC programmed with "PWM_cal." You can omit the motor driver circuit for now because it is not used in the calibration program.

(4)　Report the nominal Pot values displayed for your program for each of the active keypad buttons:

　　　pot_val for the 1-key: 100s: _____　10s: _____　1s: _____　　　value: _____

　　　pot_val for the 4-key: 100s: _____　10s: _____　1s: _____　　　value: _____

　　　pot_val for the *-key: 100s: _____　10s: _____　1s: _____　　　value: _____

(5)　Repeat Steps 1 and 2 for the "PWM.bas" program, replacing the "PWM_cal" program on your PIC. Modify the "pot_val" ranges if necessary based on the information in Step 4. Add the motor driver circuit to your board if you haven't done so already. Insert the reprogrammed PIC into your circuit.

(6)　Build and test the circuit and have your TA verify that it is functioning properly.

LAB 11 QUESTIONS

Group: _____   Names: _____

(1)   Explain how you think the Pot command works.

(2)   How do you think the motor would respond to a very low (close to 0%) duty cycle PWM signal? What affect would changing the signal frequency (f) have?

(3)   What would happen if other keys (besides the 1-key, 4-key, and *-key) are pressed down during the keypad loop? What would happen if two of the three valid keys are pressed down at once (e.g., the 1-key and the *-key)?

(4)   In the PWM.bas program, we used 30,000 microseconds for the PWM period. What frequency (in Hz) does this correspond to?

(5)   In PicBasic Pro, to what values would the following expressions evaluate?

a) 2 / 3 * 4

b) 2 * 4 / 3

Laboratory 12

Analog To Digital Conversion

Required Special Equipment:
- Computer with data acquisition (DAC) card and LabView software.

12.1 Objectives

This Lab demonstrates how to acquire data from a physical system using a personal computer equipped with a data acquisition interface card and the application LabView. The importance of the Sampling Theorem is also examined.

12.2 Introduction

A data acquisition system is used to convert an analog signal to a sequence of digital numbers that can be stored and processed on a computer. The most common type of analog signal acquired by computer is a voltage output from a sensing device. Examples are voltages due to resistance changes in a strain gage Wheatstone bridge, voltages from an accelerometer charge amplifier, and voltages from a thermocouple or thermocouple amplifier.

A data acquisition system consists of a sample/hold circuit to capture an instantaneous value of a time varying analog voltage signal, an A/D convertor to convert this voltage to a digital code, and a computer interface that allows storing and processing of the digital data. These components are packaged on an affordable PC add-on board called a Data Acquisition and Control (DAC) card. These cards support various language programming environments including C, FORTRAN, and BASIC. Various software function calls are provided via a software library which gives easy high-level access to the board's capabilities. Acquiring data from the outside world on the computer is a simple matter of calling a function from a program. A DAC card can also be controlled with LabView, a visual programming interface where icons are selected and connected to achieve the desired functionality. A DAC card can support both input and output functions including binary (TTL) I/O, analog I/O, and counter/timer features.

The A/D feature on a DAC board is usually a 12-bit converter so the resulting digital value returned is a number ranging from 0 to 4095 (2^{12}-1). The card is normally configured so that the analog input range is -5V to 5V so the smallest voltage change that the card can detect is 2.44 mV (10 V / 4095). The maximum sampling rate of a typical card is approximately 16,000 Hz. Refer to the DAC card manual for more detailed information concerning board operation, configuration, and programming.

With data acquisition, much care must be exercised in choosing a sampling rate which is fast enough to capture the analog signal's frequency content. However, too large a sampling frequency will result in an excessive and unnecessary amount of data. The Sampling Theorem states that the sampling rate must be more than twice the rate of the highest frequency component of the signal. Lower sampling rates result in aliasing, and the digital data acquired will not be a true representation of the analog signal. The effects of sampling rate will be investigated below.

12.3   Using the LabView Data Acquisition System

12.3.1   Running and Setting Up LabView

(1)   Start LabView from the Start menu (Note: LabView is loaded only on the PCs with data acquisition hardware).

(2)   Press the New VI button to bring up the two user interface windows (the "gray window" and the "white window").

(3)   In the white window, under the Help menu, select "Show Help."

(4)   In the white window, under the Windows menu, select "Show Function Palette."

(5)   In the white window, under the Windows menu, select "Show Tools Palette."

(6)   In the gray window, under the Windows menu, select "Show Controls Palette."

(7)   Move the windows so they do not overlap.

12.3.2   Adding Needed Data Acquisition User Interface Elements

(1)   Add the Analog Input block in the white window:

   - In the Functions Palette, press the Data Acquisition Button. This will bring up a submenu.

   - Press the Analog Input button to bring another submenu.

   - Press the first AI Multi Point button (AI Acquire Waveform.vi).

   - Move the cursor to the middle of the white window and place, by clicking with the mouse, the Analog Input block in the window.

(2)   Add the Array Builder in the white window.

   - In the Functions window, press the Array button.

   - Press the Build Array button.

   - Move the cursor to the white window and place the Build Array block in the window.

(3)   Add a Graph to the gray window.

   - In the controls window, press the Graph button.

150

- Press the Waveform Graph button.

- Move the cursor over the gray window and place the Graph in the window.

- This will add a small block (DBL) to the white window. Note: to move blocks in the white window, first select the arrow cursor in the Tools Palette.

12.3.3   Setting up the Data Acquisition Block

The analog input block requires four inputs: Device Number, Channel Number, number of samples, and sampling rate. They are located on the left hand side of the block. One way to set the inputs is to create constants for them. To create a constant, right click while the cursor is on the input of a block and select Create Constant. Then type the value of the constant.

For this Lab, we will use Device 1 and Channel 0. You will need to determine the number of samples and the sampling rate required for a given signal. The maximum sampling rate for National Instruments model BNC2090 DAC card is 500 kHz.

12.3.4   Wire together the User Interface Elements in the White Window

(1)   In the Tools window, select the spool of wire.

(2)   Connect the Array to the Analog Input and the Graph:

- Click on the upper right side of the Analog Input block (this is where the waveform comes out) and then click on the left side of the Array Builder. A wire should now connect the two.

- Click on the right side of the Array Builder and then click on the Graph block (DBL). A wire should now connect them.

(3)   To acquire the data, press the arrow in the grey window.

12.3.5   Connect the Hardware

(1)   Attach a coaxial cable (with BNC connectors) to ACH0 on the back of the National Instruments interface module and connect it to your signal source.

Note: the maximum voltage range for the National Instruments model BNC2090 DAC card is +/- 10V.

12.4   Laboratory Procedure / Summary Sheet

Group: _____   Names: _____

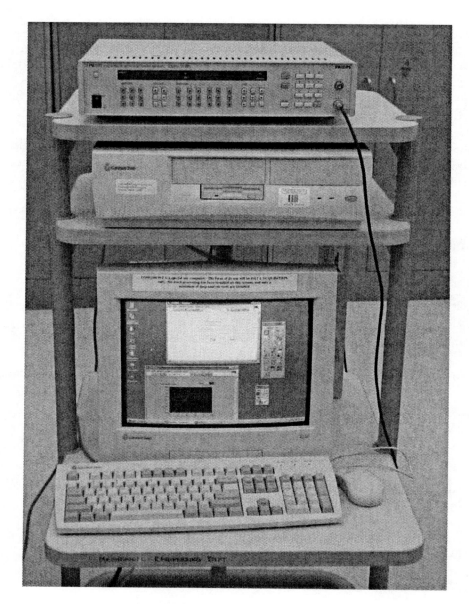

Figure 12.1 Data Acquisition System

(1)   Use the function generator to produce a 90 Hz $5V_{pp}$ sine wave so the range of voltages produced is -2.5 to 2.5 volts. Verify this signal on the oscilloscope.

(2)   Have the TA check your signal and help you attach the function generator output to the channel zero analog input of the DAC card.

(3)   Use LabView to sample five periods of this signal at the frequencies listed below. Use the hand cursor tool to change the sample rate in the Analog Input block. Record the number of data samples used for each frequency.

Sampling Rate	Number of Samples
100 Hz	
150 Hz	
175 Hz	
180 Hz	
190 Hz	
500 Hz	

(4)   Plot the results of each of these sample sets (voltage vs. time) and comment on the results and the application of the Sampling Theorem. Did you observe any aliasing?

LAB 12 QUESTIONS

Group: _____   Names: _____

(1)   Explain the results displayed on the four plots. Based on your understanding of the Sampling Theorem, why are the plots significantly different?

Laboratory 13

Strain Gages

Required Special Equipment:
- strain gage conditioner and amplifier system (Measurements Group 2120A and 2110A modules)
- strain gage interface box and cabling
- custom-made apparatus containing an aluminum tube with a strain gage Rosette mounted on its top surface.

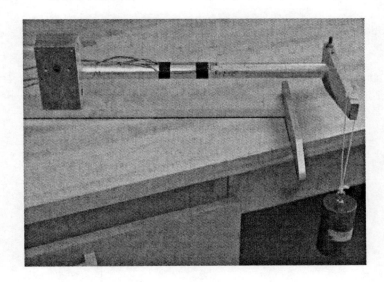

13.1 Introduction

The intent of this laboratory exercise is to familiarize the student with the use and application of strain gages. In particular, this exercise will utilize a rectangular strain gage rosette, strain gage conditioner and a voltmeter for the determination of strains within a loaded specimen. A foil strain gage and a rectangular strain gage rosette are illustrated below.

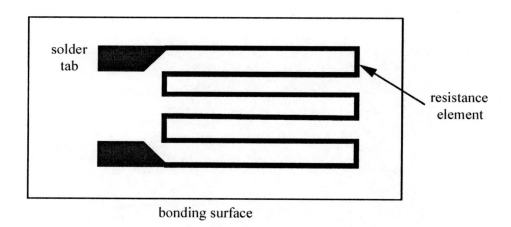

Figure 13.1 Foil Gage

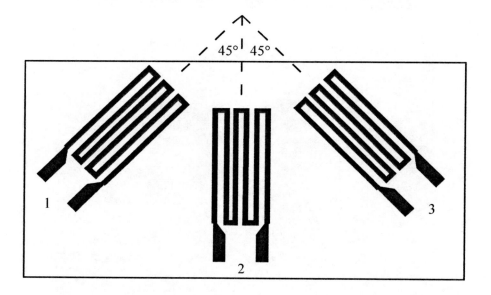

Figure 13.2 Rectangular Strain Gage Rosette

13.2   Theory

The basic principle under which the strain gage operates is the fact that the electrical resistance of a conductor changes in response to a mechanical deformation:

$$R = \rho \frac{L}{A} \qquad (11.1)$$

where

R = resistance of conductor

ρ = resistivity of material

L = length of conductor

A = cross-sectional area of conductor

Relating the above definition to Poisson's ratio and strain yields the following:

$$F = 1 + 2v + \frac{1}{\varepsilon}\frac{\Delta\rho}{\rho} \qquad (11.2)$$

$$\varepsilon = \frac{1}{F}\frac{\Delta R}{R} \qquad (11.3)$$

where

F = gage factor

v = Poisson's ratio

ε = axial strain

ΔR = change in gage resistance due to deformation

R = undeformed gage resistance

The strain gage conditioner consists of several channels, each containing a bridge/amplifier circuit. Each channel outputs a bridge detector potential, V_o, that is related to the strain in the gage connected to that channel. A bridge circuit is illustrated in Figure 13.3. For a balanced bridge (i.e., $V_o = 0$) the condition $R_1R_3 = R_2R_4$ must be satisfied. Thus, once the bridge is balanced for a no strain condition, a strain induced on the strain gage will result in a nonzero detector potential V_o. The change in this voltage can then be used to determine the corresponding strain. When the gage resistance changes, the detector voltage changes as

$$\frac{\Delta V_o}{V_e} = \frac{R_1 + \Delta R_1}{R_1 + \Delta R_1 + R_4} - \frac{R_2}{R_2 + R_3} \qquad (11.4)$$

so the change in resistance of the strain gage can be expressed as

$$\frac{\Delta R_1}{R_1} = \frac{(R_4/R_1)[\Delta V_o/V_e + R_2/(R_2 + R_3)]}{1 - \Delta V_o/V_e - R_2/(R_2 + R_3)} - 1 \qquad (11.5)$$

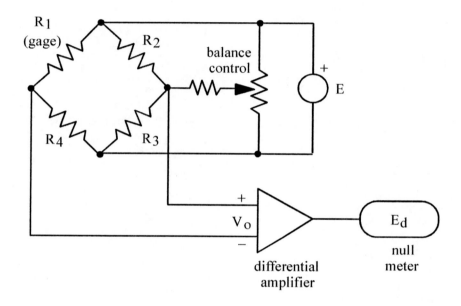

Figure 13.3 Strain Gage Conditioner Circuit

Now, turning our attention toward mechanics of materials, it can be shown that the strains ε_1, ε_2, ε_3 of the rectangular strain gage rosette are related to the principal strains and principal stresses as follows:

$$\varepsilon_a, \varepsilon_b = \frac{\varepsilon_1 + \varepsilon_3}{2} \pm \frac{1}{\sqrt{2}}[(\varepsilon_1 - \varepsilon_2)^2 + (\varepsilon_2 - \varepsilon_3)^2]^{1/2} \qquad (11.6)$$

$$\sigma_a, \sigma_b = \frac{E(\varepsilon_1 + \varepsilon_3)}{2(1 - v)} \pm \frac{E}{\sqrt{2}(1 + v)}[(\varepsilon_1 - \varepsilon_2)^2 + (\varepsilon_2 - \varepsilon_3)^2]^{1/2} \qquad (11.7)$$

and the direction of the maximum principal stress axis (σ_a axis) as measured counterclockwise from gage 1 is given by:

$$\tan 2\theta = \frac{2\varepsilon_2 - \varepsilon_1 - \varepsilon_3}{\varepsilon_1 - \varepsilon_3} \qquad (11.8)$$

and since $\varepsilon_2 > 0.5\,(\varepsilon_1 + \varepsilon_3)$, we use solution in the top half plane ($0 < 2\theta < 180°$).

The rectangular strain gage rosette senses the state of strain at a point on the top of the tubular cantilever beam. The stress due to bending at this point is:

$$\sigma_x = \frac{Mc}{I} \qquad (11.9)$$

and the shear stress is:

$$\tau_{xy} = \frac{Tc}{J} \qquad (11.10)$$

where

M = moment corresponding to applied load

T = torque corresponding to applied load

c = radius to outer surface of the tube

I = area moment of inertia of the tube

J = polar area moment of inertia of the tube

x = axial direction

The moment of inertia and polar moment of inertia for a tube are:

$$I = \frac{\pi}{64}(d_o^4 - d_i^4) \qquad (11.11)$$

$$J = 2I = \frac{\pi}{32}(d_o^4 - d_i^4) \qquad (11.12)$$

Since the shear stress on the principal planes is zero, and since σ_x is located at an angle

$$\phi = 45° - \theta \qquad (11.13)$$

from the principal axes (from Mohr's Circle), the plane stress equations give us:

$$\sigma_x = \sigma_{avg} + \frac{\sigma_a - \sigma_b}{2}\cos 2\phi \qquad (11.14)$$

$$\tau_{xy} = \frac{\sigma_a - \sigma_b}{2}\sin 2\phi \qquad (11.15)$$

where $\sigma_{avg} = 0.5\,(\sigma_a + \sigma_b)$.

13.3   Laboratory Procedure / Summary Sheet

Group: _____   Names: _____

The experimental setup is illustrated in Figure 13.4. We wish to determine the bending moment M (theoretical value = mgb), the torque T (theoretical value = mga), and the mass m by utilizing the strain gage measurements given.

<u>Properties and geometry of the aluminum tube, strain gage rosette, and hanging mass</u>:

E = 70 GPa, ν = 0.334
L = 0.395 m, a = 0.16 m, b = 0.182 m, d_o = 1.00 in, t = 0.085 in
F = 2.05
m = 1.492 kg

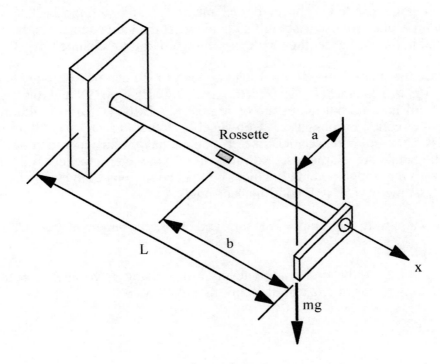

Figure 13.4 Experimental Setup

161

(1) It can be shown (see "Experimental Stress Analysis" by Dally and Riley, McGraw-Hill, 1991) that an active gage (with gage factor F) produces approximately F/4 output microvolts per microstrain and per volt excitation. NOTE - this is a unitless quantity: $\frac{\mu V}{\mu \varepsilon V}$.

Thus the equation relating strain to measured voltage is:

$$\varepsilon = \frac{V_{meas}/GAIN}{\left(\frac{F}{4}\right)V_{ex}} \qquad (11.16)$$

For an excitation potential of 5 volts, we wish to find the gain of the amplifier required to produce a 2 volt output at 500$\mu\varepsilon$. Calculate the required gain assuming a gage factor of 2.

Gain = _____

(2) The strain gage rosette is connected to channels 1, 2 and 3 on the 2120A. Make sure that the gain multiplier control is set to x200. Now set the gain control dial based on the value calculated in part 1; i.e. set the gain control to gain/200 for channels 1 – 3.

(3) Make sure that there is no external load applied to the cantilever. Now adjust the bridge balance for each channel (1 – 3); First turn the EXCIT toggle ON and rotate the BALANCE control until both output lamps are extinguished. If the (-) lamp is illuminated turn the BALANCE control clockwise. Conversely, if the (+) lamp is illuminated turn the BALANCE control counterclockwise. If you are having difficulty distinguishing whether or not the lamps are illuminated, you may use a voltmeter attached to the DAC interface card to zero the bridge potential. Under no load conditions each channel should read zero volts; adjust the BALANCE control accordingly.

(4) Hang the mass from the center of the tube and record the gage voltages. Comment on these results.

(5) Hang the mass at the end of the lever arm. Using a voltmeter, read the voltages corresponding to gages 1, 2 and 3; record them below.

V_A = _____

V_B = _____

V_C = _____

(6) Now calculate the strains in each of the three gages of the rosette; utilize the relationship from part 1.

ε_1 = _____

ε_2 = _____

ε_3 = _____

(7)   Knowing these strains determine the following:

      a) The bending moment in the beam associated with the applied load.

      b) The torque produced by the lever arm and the applied load.

      c) The mass applied at the end of the lever arm.

(8)   Submit your full analysis used to determine the value of the hung mass from the strain gage voltage measurements. Compare the calculated result to the actual value of the mass. Submit your work to your TA at the following week's Lab meeting. Comment on various possible sources for error in the measurements and analyses.

Laboratory 14

Vibration Measurement With an Accelerometer

Required Special Equipment:
- custom-made apparatus consisting of two sets of motors/shafts/bearings mounted on an aluminum plate
- Endevco 2721B charge amp
- Endevco 2256M15 or 2211E accelerometer
- high current power supply (HP 6286A)

14.1 Objective

The objective of this exercise will be to compare the vibration characteristics of a normal and a defective ball bearing turning under load. The vibrations will be sensed with two piezoelectric accelerometers. The oscilloscope spectrum analyzer will be used to compare vibrations from the two bearings to detect defects.

14.2 Background

Vibration Measurement With Piezoelectric Crystals

Piezoelectric accelerometers are in wide use for measuring shock and vibration. Most accelerometers have a design similar to that illustrated in Figure 14.1. The mass (called the seismic mass) causes inertial loads in response to motion of the object to which the accelerometer is attached. The inertial loads cause strain of the piezoelectric crystal. Due to the piezoelectric properties of the crystal, the strain causes displacement charge which is sensed at the crystal conductive coatings. A charge amplifier and conditioning circuit can measure this charge and convert it to a voltage signal which represents the acceleration of the object. A pre-loaded spring is used to keep the crystal in compression resulting in more linear behavior of the crystal.

In general, piezoelectric accelerometers cannot measure constant or slowly changing acceleration since the crystals can only measure a change in force by sensing a change in strain. But they are excellent for dynamic measurements such as vibration and impacts.

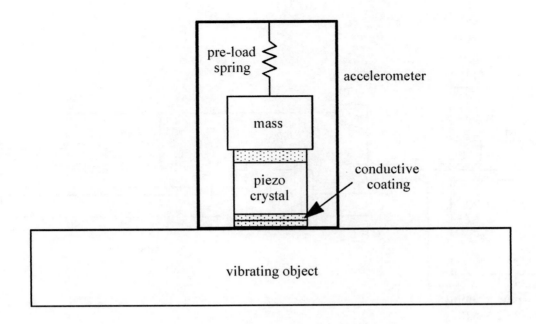

<u>Figure 14.1</u> Piezoelectric Accelerometer Construction

14.3   Theory

A defective ball bearing will cause more vibration than an undamaged bearing. We can measure bearing vibration experimentally and determine whether a bearing is defective or not. To do this we must have a clear idea what the vibration characteristics of a good bearing are. A defective bearing will have more vibration components in the high frequency range than a non-defective bearing. This is due to scratches in the balls and imbalances in the high speed rotation of the shaft.

One method to analyze the frequency components of a bearing is to record the output of the accelerometer (which is attached to the bearing pillow block) over a given period of time. The Fourier transform of this waveform will convert the vibration data from amplitude vs. time to amplitude vs. frequency.

14.4   Laboratory Apparatus

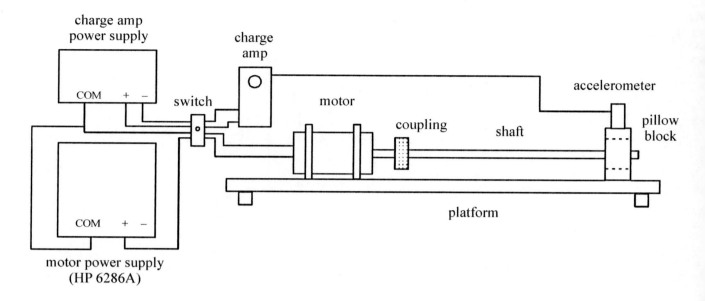

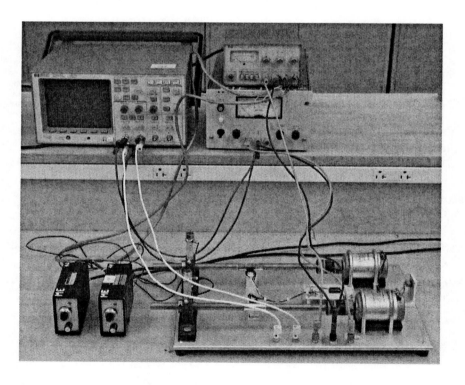

Figure 14.2   Schematic of Bearing Signature Analysis Apparatus

The apparatus in the laboratory consists of a good and a defective bearing each supporting a shaft driven by a DC motor. You will have to determine which bearing is defective after you have completed the procedures listed below.

The accelerometers are mounted on plexiglass blocks which are in turn rigidly attached to the bearing pillow block. The acclerometers are connected to charge amplifiers.

Only one motor should be operated at a time in order to prevent cross-talk of vibration between the two bearings. The supply of the corresponding charge amplifier is selected with the help of the toggle switch.

A strobe light can be used to measure the shaft speed. By drawing an axial line on one side of the shaft and a transverse line on the other side, we guarantee a reliable strobe light measurement. If the strobe is set to the shaft speed, one of the lines will be illuminated on each rotation of the shaft resulting in a stationary line image. If the strobe speed is set to half of the shaft speed, one of the lines will be illuminated on every second rotation of the shaft resulting in a dimmer image. If the strobe speed is set to twice the shaft speed, each line will be alternately illuminated on every half turn of the shaft resulting in a plus sign image. There are many more possibilities depending upon the shaft and strobe speeds, but a reliable method to acquire a good measurement is to start with a high strobe speed and decrease the speed until the image changes from a relatively bright stationary plus sign image to a bright single line stationary image.

As an alternative to the strobe light, you may use the reflective tape on the shaft and a retroreflective photosensor. The pulse train coming from the photosensor can be displaced on an oscilloscope and used to measure the frequency of revolution and thus the revolutions per second (rps) of the shaft.

14.5   Laboratory Procedure / Summary Sheet

Group: _____   Names: _____

(1)   Select the motor/bearing you want to take data from and set the toggle switch to the correct position.

(2)   Switch on the power supplies for the charge amplifier and the motors.

(3)   Set the motor speed to 3600 rpm with the aid of the strobe or retroreflective photosensor.

(4)   Look at the waveform you are getting on the oscilloscope.

(5)   Process the waveform using the Fast Fourier Transform (FFT) feature on the oscilloscope. To access this feature on an HP Digital oscilloscope, use the ± button between the channel 1 and channel 2 buttons to access a menu that allows you to perform math on the signals. Turn on Function 2 and display the FFT menu. Turn off the channel 1 and 2 displays (with the Channel buttons) so only Function 2 is on. This results in a clear line spectrum display. Sketch the vibration waveform and the FFT spectrum. Alternatively, acquire the data with the LabView DAC card and process it in MATLAB or MathCAD to generate FFT spectrum plots.

(6)   Repeat the procedure for the other motor/bearing.

Compare the two sets of waveforms and spectrum plots. Try to draw conclusions about which bearing is in better shape. Submit the sketches and comments to the TA.

Laboratory 15

A Microcontroller-based Design Project

15.1 Objective

Each group must design, build, test, and demonstrate a device controlled by a PIC microcontroller. The device must have functioning elements in six functional categories listed in Section 15.5. The device will be rated (graded) based on the level of complexity achieved in each category. There will also be grading adjustments for aesthetics, creativity, and how well the project is documented.

15.2 Project Deliverables

Each group must present the following over the course of the project:

- a proposal (see Section 15.3)

- a design notebook (see Section 15.4)

- a device containing functional elements in each of the categories listed in Section 15.5

- a final design report (see Section 15.7)

We expect each group to be creative in coming up with a unique "device" that performs some useful function. Past project and alternative ideas are displayed at *www.engr.colostate.edu/ mechatronics/projects.html*.

15.3 Proposal

The proposal must contain:

- a title page with title, group number, group member names, and date.

- a concise overview of what the proposed device is and description of how it will work. Include a well-labeled figure illustrating the concept.

- a list of proposed components in each of the functional element categories

15.4 Individual Design Notebook

The design notebook is a loose leaf binder containing notes, sketches, schematics, documents and designs. It will be separated by tabs for each week of the design cycle. Pages must be dated and initialed. The notebook will be reviewed weekly by the Lab Instructor. The Lab Instructor will record grade of acceptable or unacceptable each week.

15.5   Required Functional Element Categories

Each device must contain functioning elements in each of the six categories listed below. The examples under each element category are listed in order of increasing rating score (see Section 15.6). Other devices not listed as examples below are acceptable.

(A)   Output Display Device

- LED

- 7-segment digit display

- LCD

(B)   Audio Output Device

- buzzer

- speaker with sound effects

- speaker with digitally recorded or synthesized music or voice

(C)   Manual Data Input

- switch

- button

- potentiometer

- joystick

- keypad

- keyboard

(D)   Automatic Sensor for Data Input

- switch

- photo-optic pair

- potentiometer

- photo cell

- temperature sensor

- encoder

(E)   Actuators

- solenoid

- dc motor

- stepper motor
- pneumatic cylinder

(F)   Logic, Counting, and Control

- on-off motion control in one direction only
- counting
- programmed logic
- motion in different directions and magnitudes
- open-loop control
- menu-driven software
- closed-loop feedback control
- A/D and D/A interfaces

15.6   Functional Element Category Ratings

The group's grade for the project will be based on the device's performance in each functional element category listed above (A, B, C, D, E, F) and on several grading adjustments described below. The rating for each category will be based on the following system:

Rating	Description of performance
0	nothing implemented
2	something implemented, but non-functional
5	something implemented, but not functioning as designed
10	something discussed in class or Lab, functioning as designed (i.e., performs some intended, useful function), and is repeatable
15	something presented in the textbook, but not discussed in class or Lab, functioning as designed, and is repeatable
20	something not presented in the textbook, class, or Lab, that required independent research, functioning as designed, and is repeatable

Each category will receive a rating, and the base project score will be the sum of the six ratings. For example, if the project is rated 10 for category A, 15 for B, 10 for C, and 15 for D, 20 for E, and 15 for F, the base project score would be 85.

15.7   Final Design Report

The final report is due at the last meeting of the Lab section. The report should include:

- Title Page: title, group number, group member names, and date.

- Design Summary: concise overview of what the device does, how it works and success in meeting the original problem definition. Include a well-labeled figure illustrating the overall device.

- Design Details: detailed figures and photographs with key features and components clearly labeled, circuit schematics and/or functional diagrams and software flowcharts. The use of computer aided tools (Word, PowerPoint, PSpice, Solid Works) is encouraged. Be sure to refer to the figures and diagrams in the text and describe them completely. Include detailed wiring diagrams (if details are not included in earlier figures) and well commented software listings in an Appendix and refer to them in the body of the report.

- Design Evaluation: describe the success of the device in meeting the functional element categories and provide justifications for any anticipated grading adjustments.

After looking at the figures and schematics and after reading the BRIEF descriptions in "Design Summary" and "Design Details," the reader should be able to fully understand what the device is and how it functions (without seeing the actual device).

15.8   Grading Adjustments

The base project score (the sum of the category ratings) will be adjusted by the following grading adjustments:

- +/-5 for the proposal per the requirements listed above (-5: poor; 0: average/acceptable; 5: exceptional)

- -10 maximum for a poor design notebook (-10: many unsatisfactories; 0: completely satisfactory)

- +/-10 for the final project report per the requirements listed above (-10: poor; 0: average/acceptable; 10: exceptional)

- +/-5 based on construction quality, aesthetics, consumer appeal (e.g., perforated protoboard with neat soldered wiring vs. messy breadboard; well-built and attractive packaging) (-5: poor; 0: average; 5: exceptional)

- +/-5 based on construction cost and expected mass production cost appropriate for the functionality (-5: expensive; 0: average; 5: inexpensive)

- +/-5 for creativity, originality, and usefulness (-5: poor; 0: average; 5: exceptional)

- A group self-evaluation will occur at the middle and end of the semester. This provides an opportunity to praise and critique group members and it may be used to help adjust individual grades.

- +5 if you receive a rating of 5 or above in all six categories by the early bird date listed on the course syllabus.

NOTE - The potential for a positive adjustment increases with the level of functionality. A maximum positive adjustment (especially for the final project report) is possible only for a well-designed high-ranking device.

15.9   Additional Information

- The Lab Instructor will try to provide any common circuit components needed (resistors, capacitors, LEDs, limited ICs), and the Engineering Shop can supply you limited materials and mechanical hardware. Other items (special ICs, switches, buttons, miscellaneous mechanical and electrical accessories, etc.) must be purchased. See useful local and mail-order vendors at *www.engr.colostate.edu/~dga/vendors.html* for supplier information. The Lab Instructor may order components through lower cost mail-order vendors, but groups must submit an order sheet to the Lab Instructor. Electronic items from local retail stores can be quite expensive, so try to order all components through your Lab Instructor.

- Information and vendors for various actuators and sensors can be found at:

  www.engr.colostate.edu/mechatronics/resources.html

  Information and vendors for various I/O devices, PIC accessories, motor drivers and controllers, and sensor interfaces can be found at:

  www.engr.colostate.edu/mechatronics/pic.html

- We recommend that group members work together as much as possible, but the project work may be more manageable if tasks are divided among the group members. The entire group is still responsible for the work (e.g., if one group member doesn't do their part, the other members must take up the slack and evaluate the non contributing member accordingly). Here is an example of a list of duties to distribute among the group members:
  - project management (schedule meetings, plan and monitor progress, budget and collect for purchases, foster communication, etc.)
  - product and component research and purchasing
  - mechanical hardware design, assembly and testing
  - electronics design, assembly and testing
  - PIC microcontroller programming and interfacing
  - design documentation and report writing

  Also, the group will have to multitask, accomplishing various design and testing steps in parallel (e.g., do not wait for the microcontroller to get programmed before testing the motor, input circuits, sensors, etc.).

- Official evaluation trials will be held during the Lab section meetings during the latter part of the semester. No trials will be allowed after the last class day of the semester. Multiple trials are allowed to progressively improve the base score. A group is allowed only one official trial per day. Every group must also show their device during the last Lab section meeting for final determination of the grading adjustments.

- Selected groups will be invited to present their projects to the entire class during the last two lecture periods of the semester (see "Extra Rewards" above).

- Theoretically, the highest possible score (for an extraordinary device requiring much effort and research, functioning repeatedly, and presented well) is 155 on a scale of 100! Although such a high score (155) is unlikely, with hard work and good performance, a score over 100 can be achieved. This could help in recovering from poor grades in other parts of the course.

- The design notebook and group self-evaluation are the only evaluations considered on an individual basis. However, individual grades in the course can also be adjusted based on the group self-evaluations completed in the middle and at the end of the semester (see Course Policies).

- Group composition is very important. See Course Policies for more information about how the groups are formed and how individuals are evaluated within a group.

The remaining sections present various practical resources, considerations, and suggestions that might be helpful to you in designing and implementing your project. **Please read this material and apply the suggestions in your design**.

15.10   dc Power Supply Options for PIC Projects

There is a number of ways to provide the dc power required by the PIC and any ancillary digital integrated circuits. Actuators may also be included in the dc supply circuit if their drive voltage matches that of the digital circuitry. We begin by assuming that TTL digital ICs are used in the project, requiring a closely regulated 5V dc source. If CMOS is used exclusively, there are fewer restrictions on the regulation of the dc voltage.

Figure 15.1 shows various low cost options for powering systems requiring a 5V supply. The options include:

(1)   a 6 V, 9 V, or 12 V wall transformer with a 5 V regulator

(2)   a potted power supply with ac input and 5 V regulated output

(3)   four AA batteries (6 V) in series with a 5 V regulator

(4)   a 9 V battery with a 5 V regulator

(5)   a rechargeable battery (or batteries in series) with a 5V regulator

(6)   a full featured instrumentation power supply

wall transformer   potted   4 AA batteries   9V battery
                       power supply   in series

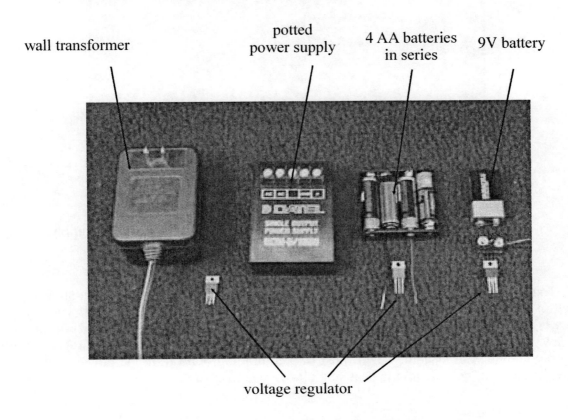

voltage regulator

Figure 15.1 Low Cost Power Supply Options

A wall transformer (6 V, 9 V, or 12 V) will provide current at its rating, and must be used with a 5V regulator to control the level of the output voltage. Be sure that the current rating of the wall transformer exceeds the maximum current your circuit and actuators will draw. A potted power supply also has ac inputs and may provide one or more regulated dc outputs at its rated current. No voltage regulator is required if a 5V output is provided. Four AA batteries may be connected in series with the 6 V output regulated down to 5 V with a voltage regulator. A 9 V battery must also be connected to a 5 V regulator. The battery options provide portability for your design but may not be able to supply enough current. Section 15.11 presents more information on different types of batteries and their characteristics. Generally, actuators such as motors and solenoids as well as LED's can draw substantial current, and batteries should be tested before assuming that they will provide sufficient current. Digital circuitry, on the other hand, usually draws very little current.

Figure 15.2 shows an example of a full-featured instrumentation power supply. This particular model (HP 6235A) is a triple-output power supply, with 3 adjustable voltage outputs, each independently current rated. A full featured instrumentation power supply provides the easiest solution, but is expensive, heavy, and generally is not portable.

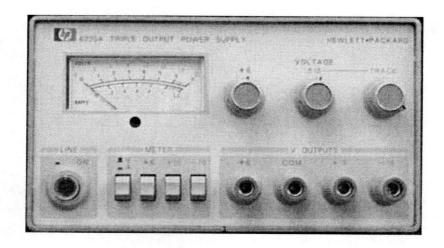

Figure 15.2 An Example of a Full-featured Instrumentation Power Supply

Except for the 5 V potted supply and the adjustable instrumentation power supply, voltage regulators are required to convert the output voltage down to the 5 V level. If your system is entirely CMOS, the regulation of the dc voltage is not required. Figure 15.3 illustrates a standard 7805 5 V voltage regulator and shows how it is properly connected to your unregulated power supply output and your system. Their must be a common ground from the power supply to your system. The mounting hole on the heat sink allows you to easily connect to the common ground.

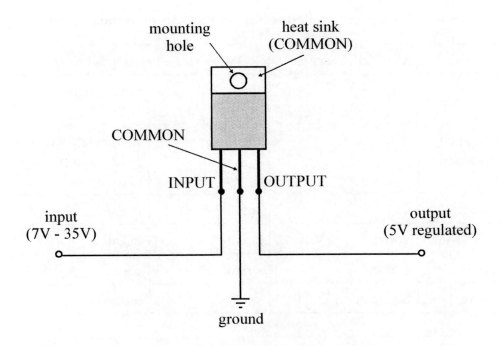

Figure 15.3 7805 Voltage Regulator Connections

Table 15.1 provides a summary of how the various power supply options compare in terms of current ratings, size, and cost. Figure 15.4 shows an example specification sheet for an enclosed power supply. Before selecting or purchasing a supply for your design, it is important to first review the specifications, especially the current rating (2.5 A in this case).

Table 15.1 5V Power Supply Options Summary

Device	Typical current	Relative size	Relative cost
instrumentation power supply	1 A – 5 A	large	very expensive (~$1000)
small potted, open frame, or enclosed power supply	1 A – 10 A	medium	moderately expensive (~$20-$100)
wall transformer	1 A	small	cheap
9V battery	100 mA	small	cheap
4 AA batteries	100 mA	small	cheap
rechargeable battery	See Section 15.11	small	moderate

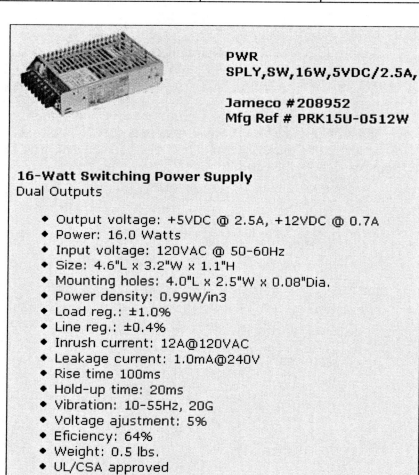

PWR
SPLY,SW,16W,5VDC/2.5A,

Jameco #208952
Mfg Ref # PRK15U-0512W

16-Watt Switching Power Supply
Dual Outputs

- Output voltage: +5VDC @ 2.5A, +12VDC @ 0.7A
- Power: 16.0 Watts
- Input voltage: 120VAC @ 50-60Hz
- Size: 4.6"L x 3.2"W x 1.1"H
- Mounting holes: 4.0"L x 2.5"W x 0.08"Dia.
- Power density: 0.99W/in3
- Load reg.: ±1.0%
- Line reg.: ±0.4%
- Inrush current: 12A@120VAC
- Leakage current: 1.0mA@240V
- Rise time 100ms
- Hold-up time: 20ms
- Vibration: 10-55Hz, 20G
- Voltage ajustment: 5%
- Eficiency: 64%
- Weight: 0.5 lbs.
- UL/CSA approved

Figure 15.4 Specifications for an Example Closed Frame Power Supply

15.11 Battery Characteristics

Many mechatronic designs will require dc voltage sources of some sort, usually tightly regulated, and often with high current capacities if actuators such as dc motors or solenoids are used. Here we present some of the important terms, considerations, and specifications in the proper selection of a battery as a power source.

The most important specification for a battery (besides its rated voltage) is the **amp-hour capacity.** It is defined as the current a battery can provide for one hour before it reaches its end-of-life point. The current that a battery can deliver is limited by its **equivalent series resistance**, which is the internal resistance that is in series with the "ideal voltage source" that is inside the battery. Batteries are composed of **cells**, the electro-chemical device that supplies the voltage and current. Cells may be combined in series or parallel within a battery for larger current and voltage capacities. The voltage of a cell will differ among the types of batteries due to their chemistry.

Primary cells are not rechargeable and are meant for one-time-use. Devices that are used infrequently or that require very low drain currents are good candidates for primary cells. **Secondary cells** are rechargeable, and their effectiveness may be replenished many times. Devices that requires daily use with higher drain currents are good candidates for secondary cells.

The plot of the **battery discharge curve** is important in determining the stability of the voltage output. Figure 15.5 shows a typical shape for a discharge curve. One desires a broad plateau characteristic for the curve.

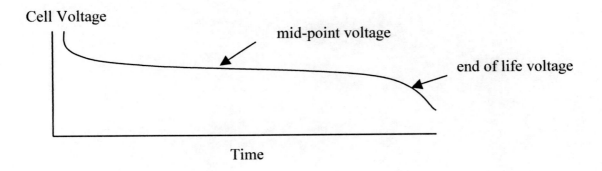

Figure 15.5 Example Battery Discharge Curve

The maximum current that a battery can deliver depends on the internal resistance of the battery. The load current times the internal resistance will result in a voltage drop reducing the effective voltage of the battery. Furthermore, there will be power dissipated by the internal resistance that, at high currents, may result in considerable heat production.

That salient factors that a designer must consider in the selection of a power source for a mechatronic design are:

o voltage required by the load
o current required by the load
o duty cycle of the system
o cost
o size and weight (specific energy)
o need for rechargeability

As shown in Table 15.2, the chemistry of the cell will determine its open circuit voltage. High drain rate devices are good candidates for lead-acid and NiCd batteries. If a device is in storage most of the time, alkaline batteries are appropriate. Since batteries may be the heaviest component of a mechatronic design, the very light Li-ion and lithium-polymer chemistries may be good candidates. Lithium chemistries provide the highest energy per unit weight (specific energy) and per volume (energy density) of all types of batteries.

Rechargeable batteries will function well even after hundreds of cycles. Rechargeable batteries are significantly more expensive than primary cell batteries. Ni-MH batteries should be deep discharged several times when put into service for best performance. Ni-Cd batteries can suffer from an effect called "memory" where the battery capacity can diminish over time. It is caused by shallow charge cycles where the battery is only partially discharged and then fully charged repeatedly. You should give the battery a deep discharge from time to time for best performance.

Table 15.2 Characteristics for Various Types of Batteries

Type	Voltage (open circuit)	Type	Typical Ah Capacity	R internal (Ω)
9V (heavy duty)	9 V	primary	0.30 @ 1 mA 0.15 @ 10 mA	35
9V alkaline	9 V	primary	0.60 @ 25 mA	2
9V lithium	9 V	primary	1.0 @ 25 mA 0.95 @ 80 mA	18
alkaline D	1.5 V	primary	17.1 @ 25 mA	0.1
alkaline C	1.5 V	primary	7.9 @ 25 mA	0.2
alkaline AA	1.5 V	primary	2.7 @ 25 mA	0.4
alkaline AAA	1.5 V	primary	1.2 @ 25 mA	0.6
BR-C PCMF-Li	3 V	primary	5.0 @ 5 mA	?
CR-V3 Mn-Li	3 V	primary	3.0 @ 100 mA	?
Ni-Cd D	1.3 V	secondary	4.0 @ 800 mA 3.5 @ 4 A	0.009
Ni-Cd 9V	8.1 V	secondary	0.1 @ 10 mA	0.84
Lead-acid D	2.0 V	secondary	2.5 @ 25 mA 2.0 @ 1 A	0.006
Ni-MH AAA	1.2 V	secondary	0.55 @ 200 mA	?
Ni-MH AA	1.2 V	secondary	1.3 @ 200 mA	?
Ni-MH C	1.2 V	secondary	3.5 @ 200 mA	?
Ni-MH D	1.2 V	secondary	7.0 @ 200 mA	?
Ni-MH 9V	8.4 V	secondary	0.13 @ 200 mA	?
ML2430 Mn-Li	3 V	secondary	0.12 @ 0.3 mA	?
Lithium Ion	3.7 V	secondary	0.76 @ 200 mA	?

15.12 Relays and Power Transistors

Actuators often require large currents at voltages different from the control circuit. Control signals are interfaced to actuator and other large current devices using relays or power transistors.

When a circuit must be completely on or off with minimal on-state voltage drop, the electromagnetic (EMR) is the only suitable choice. Solid state relays (SSRs) are the most durable and reliable but are never completely on or off and can have substantial on-state voltage drops with associated heat generation. Relays can switch dc or ac power.

Power transistors switch currents extremely fast and with less electromagnetic interference than EMRs. Power bipolar junction transistors (BJTs) and field effect transistors (FETs) can be used to switch dc power. FETs are easier to implement in a design because they do not require voltage biasing at the input. ac power cannot be switched with BJTs or FETs. Silicon controlled rectifiers (SCRs) and TRIACS are solid state devices that can switch ac power. Voltage and current capacities are important criteria when selecting any of these devices.

Here is a summary of the pros and cons of relays and transistors:

Transistors:

- can switch much faster than relays.

- produce less electromagnetic interference.

- last longer than most relays.

- can be used as current amplifiers where the output current varies with the input voltage.

Relays:

- provide electrical isolation between the signal circuit and power circuit so the control circuitry is unaffected by the power circuit.

- can switch larger currents in general.

- do not require voltage biasing at the input.

- have minimal on-state resistance and maximum off-state resistance.

- can switch dc or ac power.

15.13 Soldering

Once a prototype circuit has been tested on a breadboard, a permanent prototype can be created by soldering components and connections using a protoboard (also called a perf board, perforated board, or vector board). These boards are manufactured with a regular square matrix of holes spaced 0.1 in apart as with the insertion points in a breadboard. Unlike with the breadboard, there are no pre-wired connections between the holes. All connections must be completed with external wire and solder joints. The result is a prototype that is more robust, and that can be used in a prototype mechatronic system. You should consider this method for your class project.

For multiple versions of a prototype or production version of a circuit, a printed circuit board (PCB) is manufactured. Here, components are inserted and soldered to perforations in the board and all connections between the components are "printed" with a conducting medium. We do not support facilities to produce PCBs, but they are common in manufacturing environments.

Solder is a metallic alloy of tin, lead and other elements that has a low melting point (approximately 375°F). The solder usually is supplied in wire form often with a flux core, that facilitates melting and wetting of metallic surfaces. The solder is applied to wire and electronic components using a soldering iron consisting of a heated tip and support handle (see Figure 15.6). Sometimes you can also select the temperature of the tip using a rheostat. When using the soldering iron, be sure the tip is securely installed. Then after heating be sure the tip is clean and shiny. If not briefly wipe it on a wet sponge.

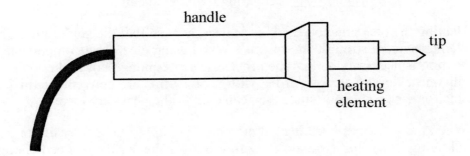

Figure 15.6 Soldering Iron

Steps in creating a good solder connection:

(1)   Before soldering, assemble your materials: a hot soldering iron, solder, components, wire, protoboard or perforated board, wet sponge and magnifying glass.

(2)   Clean any surfaces that are to be joined. You may use fine emery paper or a metal brush to remove oxide layers and dirt so that the solder may easily wet the surface. Rosin core (flux) solder will enhance the wetting process.

(3)   Make a mechanical contact between elements to be joined, either by bending or twisting, and ensure that they are secure so that they will not move when you apply the iron. Figure 15.7 illustrates two wires twisted together and a component inserted in a protoboard in preparation for soldering.

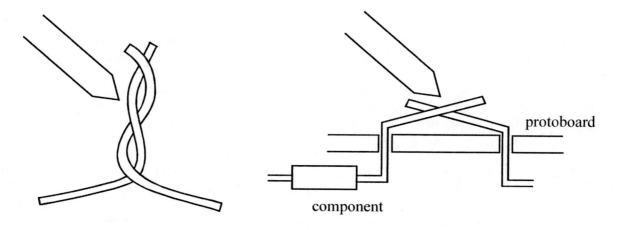

(a) wire twisted together   (b) bent leads through protoboard holes

Figure 15.7 Preparing a Soldered Joint

(4)   Heating the elements to be joined is necessary so that the solder properly wets both elements and a strong bond results. When using electronic components, practice in heating is necessary so that the process is swift enough not to thermally damage the silicon device. Soldering irons with sharper tips are convenient for joining small electronic components, since they can deliver the heat very locally.

(5)   When the work has been heated momentarily, apply the solder to the work (not the soldering iron) and it should flow fluidly over the surfaces. Feed enough solder to provide a robust but not blobby joint. (If the solder balls up on the iron the work is not hot enough.)   Smoothly remove the iron and allow the joint to solidify momentarily. You should see a slight change in surface texture of the solder when it solidifies. If the joint is ragged or dull you may have a cold joint, one where the solder has not properly wetted the elements.   Such a joint will create problems in conductivity and must be repaired by resoldering. Figure 15.8 illustrates a successful solder joint where the solder has wet both surfaces, in this case a component lead in a metal hole perforated board.

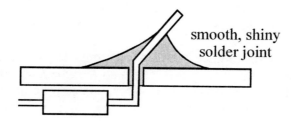

Figure 15.8 A Successful Solder Joint

(6)   If flux solvent is available, wipe the joint clean.

(7)   Inspect your work with a magnifying glass to see that the joint has been properly

made.

Often you may have a small component or integrated circuit (IC) that you do not want to heat excessively. To avoid excessive heat with a small component, you may use a heat sink. A heat sink is a piece of metal like an alligator clip connected to the wire between the component and the connection to help absorb some of the heat that would be conducted to the component. However if the heat sink is too close to the connection it will be hard to heat the wires. When using an IC, a socket can be soldered into the protoboard first, and then the IC inserted, thereby avoiding any thermal stress on the IC.

When using hook-up wire, be sure to use solid wire on a protoboard since it will be easy to manipulate and join. Wire must be stripped of its insulating cover before soldering. When using hook-up wire in a circuit, tinning the wire first (covering the end with a thin layer of solder) facilitates the joining process.

Often you may make mistakes in attaching components and need to remove one or more soldered joints. A solder sucker makes this a lot easier. To use a solder sucker (see Figure 15.9), cock it first, heat the joint with the soldering iron, then trigger the solder sucker to absorb the molten solder. Then the components can easily be removed since very little solder will be left to hold them.

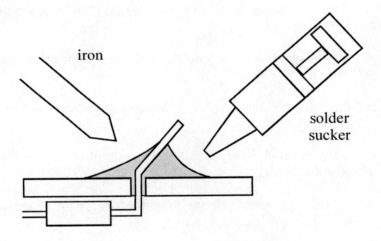

Figure 15.9 Removing a Soldered Joint

15.14 Other Practical Considerations

Here are some other practical suggestions for microcontroller-based designs:

General Electrical Design Suggestions:

- Make sure your power supply can provide adequate current for the entire design. If necessary, use separate power supplies for your signal and power circuits.

- Use breadboards with caution because connections can be unreliable and the base plate

185

adds capacitance to your circuits. Hard-wired and soldered protoboards are much more reliable. Be sure to use sockets for all ICs on the protoboard to prevent damage during soldering and to allow easy replacement of the ICs.

- Use a storage capacitor (e.g., 100 μf) across the main power and ground lines of a power supply that does not have built in output capacitance (e.g., batteries or a wall transformer) to minimize voltage drops during output current peaks. Also, use bypass capacitors (e.g., 0.1 μf) across the power and grounds lines of all individual ICs to suppress any voltage spikes.

- Avoid grounding problems and electromagnetic interference (EMI). Section 2.10 in the textbook presents various methods to reduce EMI, specifically using opto-isolators, single point grounding, ground planes, coaxial or twisted pair cables, and bypass capacitors.

- Don't leave IC pins floating (especially with CMOS devices). In other words, connect all functional pins to signals or power or ground. As an example, do not assume that leaving a microcontroller's reset pin disconnected will keep a microcontroller from resetting itself. You should connect the reset pin to 5V for an active-low reset or ground for an active-high reset, and not leave the pin floating where its state can be uncertain.

- Be aware of possible switch bounce in your digital circuits and add debounce circuits or software to eliminate the bounce.

- Use flyback diodes on motors, solenoids, and other high inductance devices that are being switched.

- Use buffers, line drivers, and inverters where current demand is large for a digital output.

- Use Schmitt triggers on all noisy digital sensor outputs (e.g., a Hall-effect proximity sensor or photo-interrupter).

- Use a common-emitter configuration with transistors (i.e., put the load on the high side) to avoid voltage biasing difficulties.

- Be careful to identify and properly interface any open-collector or open-drain outputs on digital ICs (e.g., pin RA4 on the PIC).

- For reversible dc motors, use "off-the-shelf" commercially available H-bridge drivers (e.g., National Semiconductor's LMD 18200).

PIC-related Suggestions:

- Follow the microcontroller design procedure in Section 7.9 of the textbook.

- Modularize your software and independently develop and test each module (i.e., don't write the entire program at once expecting it to work).

- Use LEDs to indicate status and location within your program when it is running, and to indicate input and output states.

- Be aware of the different characteristics of the I/O pins on the PIC. Refer to Figures 7.15 and 7.16 in the textbook to see how to properly interface to the different pins for different purposes.

- Be aware that PicBasic Pro commands totally occupy the processor while they are running (e.g., the line after a SOUND command is not reached or processed until the SOUND command has terminated).

- Refer to Design Example 7.1 in the textbook for ideas on how to interface to 7-segment digital displays with a minimum number of pins.

- When prototyping, use a socket to allow easy installation and removal of the PIC without damaging its pins.